D0203778

FIRST SIGNALS

*John Tyler Bonner*

# *First Signals*

## The Evolution of Multicellular Development

PRINCETON UNIVERSITY PRESS
PRINCETON AND OXFORD

Published by Princeton University Press, 41 William Street,
Princeton, New Jersey 08540
In the United Kingdom: Princeton University Press, 3 Market Place,
Woodstock, Oxfordshire OX20 1SY



*Library of Congress Cataloging-in-Publication Data*

Bonner, John Tyler.
First signals : the evolution of multicellular development /
John Tyler Bonner.
p. cm.
Includes bibliographical references (p. ).
ISBN 0-691-07037-7 (alk. paper)—
ISBN 0-691-07038-5 (pbk. : alk. paper)
1. Developmental biology. 2. Developmental cytology.
3. Cells—Evolution. 4. Cell interaction. I. Title.
QH491 .B595 2001
571.8′35—dc21 00-039976

This book has been composed in New Baskerville and
Bulmer Typefaces

The paper used in this publication meets the minimum
requirements of ANSI/NISO Z39.48-1992 (R1997)
(*Permanence of Paper*)

www.pup.princeton.edu

Printed in the United States of America

10 9 8 7 6 5 4 3 2 1

10 9 8 7 6 5 4 3 2 1
(Pbk.)

*To Rebecca, Jonathan, Jeremy, and Andrew*

## CONTENTS

# PREFACE

INITIALLY, I wanted to sort out in my own mind the relationship between what was developmental biology before the molecular revolution in the middle of the century, and what it is today. I wanted to find a way of looking at the goals and the achievements of both eras to see to what degree they are the same and to what degree they differ. My desire was to do more than just summarize the two; I wanted a new way of trying to find a common ground. Perhaps my twin interests in life cycles and cellular slime molds could be put to good use.

The idea was to look at a defining moment in the evolution of organisms where development could be examined from the perspectives of both pre- and postmolecular developmental biology. The moment I chose was the beginning of multicellularity, where signaling between cells began and development was at its simplest. Looking at the origin of multicellularity might be a means of isolating and revealing the universal elements of development.

To do this meant considering a number of related topics: multicellularity arose numerous times and this fact is reflected in a wide variety of present-day primitive organisms. Size and its natural selection plays an important role in those early beginnings. The invention of sending and receiving signals between the early cells is a window to understanding the development of all multicellular organisms. Examining the origin of multicellularity helps us to identify the basic elements of all of development, such as polarity and pattern formation.

There are three ways of looking at the origin of multicellular development: the straightforward, descriptive biological way; the way of molecular biology; and the use of mathematical models to search for insights. I will apply all three of these to what is known of the development of cellular slime molds. By looking at one relatively primitive organism in these three ways it might be possible to see the interplay between complexity and simplicity in development. Through such a search it might be possible to understand the essence of development.

THIS book has been especially slow in coming into full bloom, and there are many people whom I want to thank. I begin with Donna Bozzone for giving me the initial urging to write a book about the way developmental biology has changed during the course of my career. Emily Wilkinson gave me the next push; she thought I should go back and think about some of the problems of development that I raised in my first book, *Morphogenesis* (1952). Next, an invitation from Vidya Nanjundiah to a conference on development in a glorious spot in the foothills of the Himalayas helped me to gather some of my scattered thoughts and begin in earnest.

Some time ago, I finished a draft of this book, and because it was so rough the mere thought of it fills me with embarrassment. To my great benefit, it was read by Edward C. Cox, Henry Horn, Laura Katz, Evelyn Fox Keller, and

Jonathan Weiner, and my deep thanks to them all for their invaluable and wise help. No one told me to give up—in fact, some were positively encouraging—but even in the kindest comments I sensed considerable unease with what I had written. My message was not clear, my foray into philosophy was not a success (it has been tossed out!), I was even gently chided for failing to control my inner compulsion to finish things—that I should take some time to carefully rethink and crystallize my argument.

The next round was a more focused book, but still not quite there. In the final draft I was greatly helped by Dan McShea, again Jonathan Weiner, two anonymous readers, and some helpful advice from Mary Jane West-Eberhard. I shudder to think where I would be without their criticisms, constructive and otherwise.

I acknowledge with thanks Peter Karieva and *Integrative Biology* for help with an earlier version of what is to be found in chapter 3, and for Laura Katz for her assistance with figure 1 in that chapter. I thank all the kind people in the Princeton University Biology Library for their invaluable help. Finally, I would like to thank Sam Elworthy for his interest and help with the book, and Alice Calaprice for her masterful editing, a wonderful gift that has again been used to my great benefit.

I dedicate this book to our young: Rebecca, Jonathan, Jeremy, Andrew, and their families, for all they have done for me.

*Margaree Harbour*
*Cape Breton, Nova Scotia*

# FIRST SIGNALS

# 1 *Introduction*

SINCE the 1940s, I have been thinking about how animals and plants and other organisms develop. As I look back over the many years, I see that the problem remains in some ways much the same, but in other ways it has changed to an extraordinary extent. The latter is entirely due to the fact that there has been such vast progress in the very nature of biology due to the eruption of molecular biology at midcentury. There has been a certain amount of tension between classical developmental biology and its modern molecular form, largely because the techniques are so different. But that is slowly changing with the present-day realization that the molecular developmental biologist is in fact pursuing the same goals that were first staked out by embryologists in the nineteenth century. They are pursuing two different ways of looking at the same problem; they are not opposed—they complement one another.

The premolecular biologists sought the immediate causes of the steps of development, the sequences of stimuli and responses that were responsible for the evolving of the embryo. They fully realized that ultimately those stimuli and responses were chemical in nature, but it seemed beyond the abilities of the times to find out what those chemicals might be. To give an example, in the early 1900s

Hans Spemann and Hilda Mangold (see Hamburger, 1988) discovered that the dorsal lip of the blastopore could, when transplanted to another embryo, cause the surrounding tissue to produce an extra embryonic axis. They called the dorsal lip region an "organizer," fully realizing the obvious fact that it sent out a stimulus and that the response to that stimulus was the formation of a second embryo, with all its complex tissues. There followed many years of research in many distinguished laboratories to find what chemical substances might be responsible for this "embryonic induction." The results were quite confusing and unsatisfactory, and it was not until the eclosion of molecular biology that it became possible to begin to analyze the molecular basis of Spemann and Mangold's experiment. Now we know many of the substances involved—not only the specific proteins, but the genes that are responsible for their production. In fact, the level of detail of our present knowledge is quite staggering. But today those molecular details can easily cloud the underlying fundamental biological questions.

There are presently numerous and rapidly evolving reviews and books of this wonderful progress in molecular development, and it is not my intention to pursue here this exciting field of knowledge. Rather, I want to take a different approach.

In the case of the molecular biology of development, the explanation of even the smallest step has become enormously complex, a complexity that increases directly with the zest for digging. I am not criticizing this approach; in fact, I think it is essential and important and has produced stunning advances. However, I worry that all the wonderful detail will make the fundamental principles harder to see, and that perhaps one might also get illuminating insights by examining the development of simpler organisms—by which I mean simpler than those workhorses of molecular biology, the fly and the worm. Of course, what one finds is that even with "simple" organisms, the more one delves into their developmental mechanisms, the more complexity emerges; the difference is only a matter of degree.

How can one simplify the mechanism of development? How can one cut through the details and get to the core of how organisms develop? One might argue that the details *are* development, but I will take a different tack. It seems to me there are some basic principles that underlie all the details. It is upon them that the details rest, and this essay is about how to find those principles and what they might be.

Here I will consider three ways one could pursue this quest for simplicity. One is the straightforward, descriptive biological approach where the mechanisms are at least superficially exposed.

Another way is to look for the beginnings of multicellular development, where one might assume that at first only the minimum steps necessary were present. What happens inside a cell is incredibly complex despite its small size, but with the evolutionary origin of multicellularity there must have occurred some minimal signals between cells, and that was the origin of multicellular development. Those extracellular beginnings were simple and were only subsequently followed by an increase in complexity.

The third way to seek simplicity is through mathematical modeling. One can ask what is the simplest way to achieve some developmental change in form. This is an approach to which I shall return periodically to see how it fits in with my more central message of what might have been the nature of the first cell signaling when multicellular development arose in early evolution.

I do not mean to imply that these are the only ways of achieving simplicity, for there are others as well. To give a well-known example, the foundations of molecular genetics are achieved by Max Delbrück and Salvador Luria using viruses (bacteriophages), and by the very simplicity of these naked, parasitic genes (and their rapid rate of reproduction) it was possible to gain an extraordinarily deep understanding into the fundamental nature of mutations. The use of viruses to solve important biological problems remains important today, as shown recently by Birch and Chao (1999), who have elucidated some basic questions of population genetics, again by taking advantage of the molecular simplicity of viruses.

As indicated above, I want to use a different "simplicity" approach here. My plan is to get at the essentials of developmental biology by piecing together how the development of multicellular organisms might have arisen in the first place. This will be my window to finding the underlying core of development without interference from the obscuring details. But another equally important evolutionary theme runs in parallel. There is a constant selection pressure for size change: this is a major driving force in evolution. In a complex environment such as exists today for every category of organism, a complete array of size niches is present, and should any one be vacant, there will be a selection for a slightly larger or slightly smaller species or variants to fill it. In early geological times, when the world consisted solely of single-cell organisms, the selection pressure for size increase must have been great, and the easiest way to achieve it was by becoming multicellular. The unicellular size niche was filled; expansion could only be in the larger direction. This explains why there were so many independent (or convergent) inventions of multicellularity.

What we have, then, is the evolutionary origin of multicellular development. First, there was a selection for an increase in size by becoming multicellular, and once achieved there was a selection for a better integration, a better coordination of the adhering cells to compete effectively for energy and for a way to reproduce successfully. Then, with each successive step of size increase, propelled always by the fact that the uppermost size niche is never filled, there has been a further selection pressure for integration and coordination, often by new and innovative devices to accommodate the newly created larger organism.

We will examine these themes in the following way. After a bit of history of developmental biology as a discipline (chapter 2), our starting point will be a natural history of present-day simple multicellular organisms to provide a basis for speculation of how they might have arisen in the first place (chapter 3). Next, we must look into the matter of size and size niches, and in particular how it affects microorganisms (chapter 4). What are the physical properties of small clumps of cells? There are the chemical properties to consider as well: how can the cells start to communicate with one another, and what are the beginnings of the signal-response system (chapter 5)? This leads directly into the central chapter of the book, which examines the beginnings of multicellular development as a way to reveal the simplest elements of development (chapter 6). I use the cellular slime molds to show how the development of one organism can be analyzed from a classical developmental point of view and from the point of view of molecular development, and see to what extent the mathematical modeling of development might be helpful (chapter 7).

When I first began thinking about these problems, I imagined myself bucking the trend, for so many people are working on the mathematics of "complexity theory" these days; instead I was after "simplicity theory." But clearly our goals are the same, and the point of complexity theory is to simplify, to untangle the tangle. The complexity people are grappling directly with the complexity, while I am doing an end run by looking at the underlying problem before it becomes complex.

# 2 *From Embryology to Developmental Biology*

DEVELOPMENTAL biology started as a descriptive science but progressively became more experimental. Karl Ernst von Baer, in the early 1800s, was one of a long series of descriptive embryologists whose lineage went back to Aristotle; he was the first to see and describe the mammalian egg as well as the fundamental germ layers of embryos. But later in the century, the experimental approach took hold with the wonderful work of Wilhelm Roux, Hans Dreisch, and numerous others. When I was a student in the late thirties and early forties, this was almost the only way to study development. (I will return to the "almost" presently.) We had to read the endlessly long classical papers (all in German!) which, despite their difficulty, seemed at the time modern and exciting, even though by then much of the work was already antique. Animal embryology reigned supreme; it was the way to study development. Not only were amphibian embryos the objects of the experiments, but so were other vertebrates such as the chick and many invertebrates, in particular the sea urchin. The progress well into the middle of this century was enormous, and what could be learned from cutting and grafting experiments seemed endlessly fascinating. As a beginning student, my favorite text was *The Elements of Experimental Biology* by J. S. Huxley and G. R. de Beer, which had been published in 1934. I thought of it then as a beacon, a guide to a bright future.

However, there were some early exceptions to the apparent hegemony of animal embryos. In botany, the illustrious nineteenth-century Germans, such as Julius Sachs (1882) and Karl Goebel (1900) and many others, also made great strides in understanding how plants develop. It was not just the process of fertilization and the development of the early embryo that received attention, but the realization that a large plant had a "continuing embryology"; a tree is an embryo all its life—it does not stop developing in the manner of an animal embryo turning into a static adult. Furthermore, higher plants, unlike animal embryos, have stiff cell walls, which makes for interesting differences in their morphogenesis. While a considerable amount of experimental work was done on plants in the nineteenth century (including some by Charles Darwin), the big advances began with the ingenious experiments of Fritz Went in the 1930s (see Went and Thimann, 1937). He showed that the growth of plants was governed by a growth hormone (auxin), and this led to a great wave of important studies on the development of higher plants. To me, as a young student, it seemed quite remarkable that those working with animals payed so little attention to plants, and the converse could be said of those working with plants.

After the Second World War, when I was finishing my graduate studies, two friends and I organized a seminar in which Edmund R. Brill represented animal embryology,

William P. Jacobs higher plant development, and I was in charge of the development of "lower forms." The point of our joint effort was to see what we had in common, what were the broad general principles. A few years later I wrote up my version of the fruits of our labor in a book, but I am getting ahead of my story.

There was an important aspect of experimental developmental biology that had an ancient history: the study of regeneration. It is different from normal development in that to show regeneration an experiment must be performed, either by nature or by a person. The ancient Greeks knew, and no doubt it was known long before, that plant cuttings could regenerate, and this prevailed as a commonplace understanding. In the eighteenth century, Abraham Trembley, in a wonderful treatise (1744), showed that the freshwater polyp *Hydra* had quite extraordinary powers of regeneration. He worked with *H. viridis*, a species green in color, and he declared that his initial motivation for doing the experiments was to see if *Hydra* was a plant or an animal. Since it could regenerate, surely it had some plantlike characteristics, but then he found it moved like an animal. Now we know that it is an animal, a hydroid, which is a cnidarian, and that its green color is due to a unicellular photosynthetic alga living in the cells.

Hydroid regeneration continued to be vigorously pursued. In the nineteenth century, Hans Driesch (reviewed by him in 1908) and others, working in the Naples marine biological station, found the hydroid *Tubularia*, with its long stem, to be particularly suited to regeneration studies. Later I will discuss this work in some detail, but for over a hundred years *Tubularia, Hydra,* and other hydroids have made important contributions to the study of development.

Since Trembley's time, the experimental study of regeneration has continued to be a field that paralleled embryology. One aspect that received special attention over the years has been limb regeneration—of arthropods, newts, salamanders, and other vertebrates. All practitioners were zoologists and considered themselves essentially embryologists—they were embryologists who did not necessarily work with embryos. So in this respect adult animals are embryos too.

But the botanists still lived in another country and the communication between the two was nonexistent. Many universities had separate zoology and botany departments—and some still do today—clinging to them with amazing vigor. This was partly true because the botanists were definitely in the minority in our anthropocentric world, and they felt any kind of merger would swamp them out of existence. The oneness of all living things so strongly advocated many years earlier by Thomas Henry Huxley in the form of "biology" was very slow in coming into being. There were a number of reasons for the rapprochement between botany and zoology: it became evident that the study of genetics, evolution, and development had to transcend any such barrier.

The study of evolution in general took a big leap with the advent of the modern synthesis, which began with the population genetics of R. A. Fisher, J.B.S. Haldane, and Sewell Wright (see Provine, 1971, for a history of this era). It paved the way and gave an increasing sense that their principles applied to all organisms. Darwin certainly understood this, and indeed he made major contributions to botany as well as zoology, but the population geneticists of the 1930s were zoologists. The person who broadened the perspective was G. Ledyard Stebbins when he published his *Variation and Evolution in Plants* in 1950. Suddenly plants got equal time, and they were clearly fascinating and important in their own right. There were many who had appreciated this all along, for it seemed self-evident; but now it became, and remains today, a universally held commonplace. The fact that, in general, animals are motile and plants sedentary has raised interesting and revealing contrasts in how they evolve.

The expansion of embryology into developmental biology, which included regeneration and the period of size increase in the life cycle of all organisms, was a gradual process, and by midcentury it was all but complete. It was unofficially recognized by the very use of the term "developmental biology," which gradually came to supplant the more restricted "embryology" of animals and higher plants. The change was speeded by respected embryologists such as Paul Weiss (1939) who championed the new term.

It was at the onset of this era that I began as a biologist. At first I was interested in all of biology, but during my freshman year at Harvard I came under the spell of Professor William H. Weston, who taught the first half of the "botany" course on lower plants, or cryptogamic botany as it was called then. His wonderful enthusiasm for the fascination of bacteria, algae, fungi, and various kinds of slime molds totally infected me. He not only made them come alive as organisms, but he was especially interested in learning about the details of their life histories by means of experiments. In fact a large number of graduate students received their degrees under him while working on fungi (especially water molds) and slime molds, which at that time were considered fungi or at least closely allied to them. I wanted to become a cryptogamic botanist—but I had not completely grown up yet! In the next few years I was exposed to embryology with Professor Leigh Hoadley, first the basic course, and then a graduate course, where I was the overwhelmed undergraduate. It was there that I began reading Spemann and Roux and others and became totally absorbed by *Entwicklungsmechanik,* or "developmental mechanics," and "causal embryology." Now I had two absorbing interests, but I could not simultaneously be a cryptogamic botanist and an animal embryologist. Suddenly one day, the obvious solution hit me: I could use lower plants to attack the problems of embryology. Unwittingly, I was becoming part of the new movement towards developmental biology. And I had solved my own problem by merging my interests.

This is where I was when I began the seminar with Edmund Brill and William Jacobs, and was able to give full cry to my enthusiasm for the merger in 1952 when I published my first book, *Morphogenesis.* It fit in with a growing trend of interest in using bacteria, fungi, algae, protozoa, sponges, and other invertebrates for experiments in developmental biology. After the book was published, people would ask me where they could learn more about these interesting lower forms, and I would recommend all sorts of other books, ones that were not centered on development but on the natural history of the organisms themselves (for example, Smith, 1955, and Hyman, 1940). Some of them are still good reading—they refuse to go seriously out-of-date. Since for simple organisms their life cycle is their development, the two stand in close relation to each other. But the embryologists of old did not know they existed, and the cryptogamic botanists had not worried much about the problems of development.

## What We Want to Know about Development

Developmental biology has gone through the same stages as the other branches of biology: first there was description, exemplified by von Baer's discovery of the germ layers, which was soon followed by a flood of experiments to understand how one step in development led to the next; that is, what were the immediate causes. This pursuit has led to an ever-increasing successful and rewarding spiral into reductionism, so that now molecular developmental genetics has the center of the stage. We are being showered with an extraordinary wealth of "morphogens," substances that are key signaling molecules in the sequential steps of development. We know more and more about the genes that are responsible for the production of the morphogens, and the receptors for those morphogens, and the genes for the receptors. It is like watching an ever-increasing telephone book enlarge. Furthermore, there is a separate expanding compendium for mammals, fish (zebra fish), fruit flies *(Drosophila)*, nematode worms *(Caenorhabditis)*, slime molds *(Dictyostelium and Polysphondylium)*, and some fungi (in particular yeasts), algae (*Chlamydomonas and Volvox*), myxobacteria *(Myxococcus),* and others. Important advances have also been made in the developmental genetics of higher plants (*Aribidopsis*).

## Seeking Simple Explanations of Development

This great flush of reductionist triumph naturally leads one to ask whether there is an overriding macro-explanation that we should look for before we become smothered by all the detail, as wonderful as it is. At first glance it would seem that there is no very new or exciting answer to this question, and that we must settle for the traditional explanation that there is a cascade of sequential causations, which brings us with a rush into the details of all those steps. This may be as far as we can reach, but I would like to think otherwise. There is a yearning for simple explanations that transcend the details, even though they are utterly dependent on the latter.

One successful reach for simplicity may be found in the application of mathematics to the problem of development. A pioneer was Alan Turing (1952), who showed that developing pattern in organisms could be described by reaction-diffusion equations. This was an elegant way of demonstrating that theoretically many patterns that arose during development could be explained by a very simple mechanism. The requirements were (1) an activator substance that diffused and activated itself—it was autocatalytic; (2) an inhibitor of the activator that diffused more rapidly, that is, a smaller molecule. Once these two substances were set up in a gradient, they were capable of producing a pattern along that gradient.

This mathematical statement showed in a flash how chemical substances—which he called "morphogens"—could by their diffusion and their reactions account for the developing pattern. For this kind of model there was no need to know the exact chemical nature of the morphogens and how they are synthesized; the importance of the models is to show how, in principle, pattern and shape can be achieved. Over the intervening years there have been many refinements and modifications of the Turing explanation, but the basic principle is immutable. It is interesting that today there is a big movement among some developmental biologists to bring together the models with the molecular details; it is an attempt to merge the micro and the macro-explanations. This approach is not universally popular because many molecular developmental biologists feel that the be-all and end-all are the molecules themselves, and a mathematical model is hokus-pokery, quite irrelevant to the micro-explanation grail. There are others, myself included, who feel that the Turing mathematics, and all that has followed, has been a great boon, for it tells us that simple explanations are possible and therefore serve as a guide, a beacon for our search. The models are a perfect example of how an overarching explanation can gather together and organize all the pieces of the micro-explanations.

In my own evolution, there are two ways I have gone about this quest for simplicity. One is to concentrate on the development of simple organisms, and much of what is to follow will consider a variety of those life forms. In some ways this has been a useful and fruitful approach, although the hoped-for simplicity has been elusive. The reason for that is quite obvious: a single cell, either that of a prokaryote or a eukaryote, is in itself so incredibly complicated that even when such cells come together in a small multicellular unit, their innocence was already lost before they started to join together. Nevertheless, we can learn many things from the origin of multicellularity, an exercise that puts us in a good position to examine how multicellular development first arose.

# 3 *The Origin of Multicellularity*

THE APPEARANCE of multicellularity during the course of early evolution is one of the major transitions during the whole span of biological evolution, as Maynard Smith and Szathmary (1995) and others have pointed out. These transitions are especially important in their implications for natural selection because with each transition one moves from one level of selection to another. This is the case with the invention of multicellularity, where one shifts from the cell as a unit of selection to a multicellular group of cells as a unit. Cells will either compete with one another or cooperate, and it is only as they shift from competition to cooperation that they can rise to the higher multicellular level of selection (Michod, 1999; Buss, 1999). Once the cells are grouped together in harmony, they as a group begin to compete with other multicellular groups. This internal harmony may only be partial, for the cells may continue to compete, among themselves, but within the rules set by the higher multicellular level. The same applies for all the other levels of selection: the organelles within the cells and among the genes themselves continue to compete, as Dawkins (1976) has emphasized. This is also true for levels above that of multicellularity, and later I will have occasion to mention social insects as examples of such a transition. The whole colony becomes a higher unit of selection (as Darwin himself foresaw) and the individual insects show a remarkable degree of cooperation. All this is the genetic view of these evolutionary transitions, but there are other views as well. The one I now want to discuss is that grouping together automatically means an increase in size, and this has many consequences.

It is quite easy to postulate what might be advantages for organisms to become larger by becoming multicellular. For instance, Gerhart and Kirschner (1997) have argued that a multicellular organism has gained the advantage over a unicellular ancestor because it can more effectively shield itself from the vagaries of the environment by producing its own internal environment. In broader terms, this is Dawkins's (1976) argument that a competitively effective way of carrying the genes from one generation to the next is by building a complex soma that safely sees to it that the germ plasm survives.

My line of reasoning, which will be developed further presently, is that the first step in the evolution of multicellularity was a size increase due to an accident, for instance a mutation that prevents the daughter cells from separating. If the larger cell mass has any advantages, such as ensuring the safety of the genes from one generation to the next by producing a protected internal environment, then natural selection will see to it that the novelty is retained. I would argue then that size increase came first, and the possible advantages that this change might provide would follow.

Many of the recent discussions of the appearance of multicellularity are confined to animal evolution. It is usually inferred that this was a unique event that did not spawn appreciable diversity until the great proliferation of body types in the Precambrian era. This view has arisen partly because of the fossil record, which had been notably sparse until the rich findings in the relatively late Burgess Shale and similar deposits. But as someone who was raised as a cryptogamic botanist—whose traditional interests lie with bacteria, algae, fungi, and slime molds—it has always seemed to me that this stress on animal evolution is rather anthropocentric. If one thinks in terms of all organisms known to exist today, or to have existed in the past, then one can only be impressed by the diversity of the beginnings of multicellularity itself.

## A Brief Description of the Major Origins of Multicellularity

The danger in describing all the different steps towards multicellularity is that the reader could well become overwhelmed by the detail. My other difficulty is that it is a subject I have already discussed in some form or other previously (1974, 1988, 1996). Here I want to do something different. To begin, I will modify one of the current molecular phylogenies of the major groups of organisms as proposed by Sogin (1991) based on the analysis of a small subunit of the ribosomal RNA gene (fig. 1). While the use of additional markers could change the specific relations

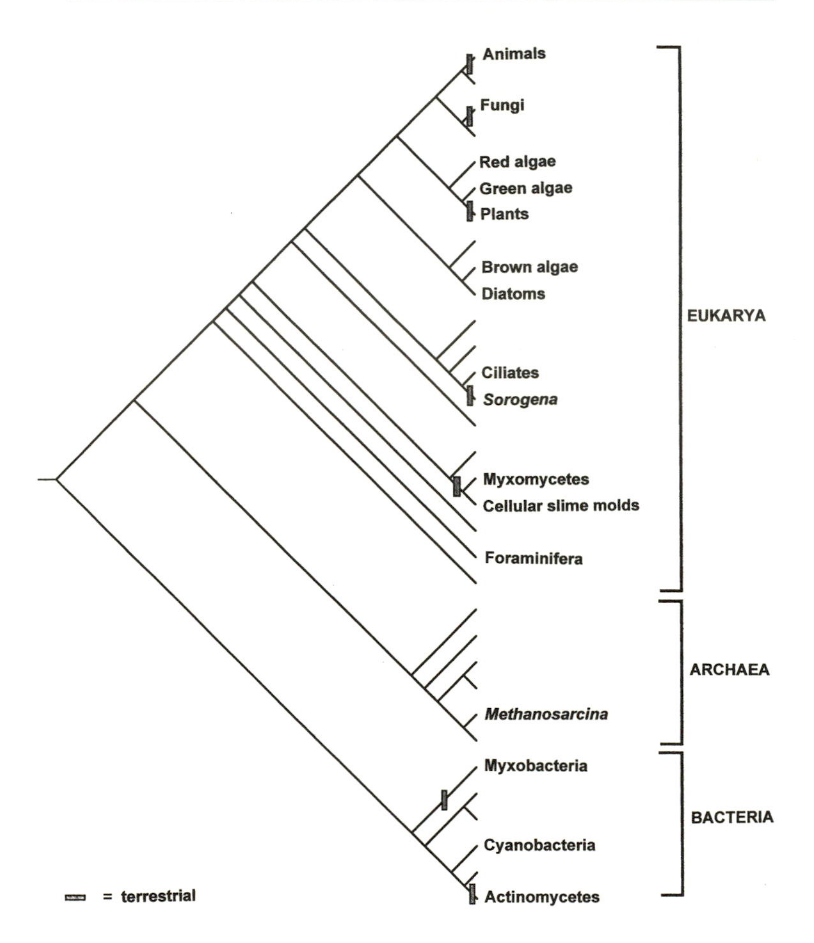


Fig. 1. A molecular phylogeny of the major groups of organisms, showing that multicellularity (the labeled branches) evolved independently a number of times. The tree is based on that of Sogin (1991), who used a small subunit of the ribosomal RNA. Information has been added from other molecules, such as certain proteins that alter some of the relations of the various taxa (see Baldauf, 1999). The rectangles indicate terrestrial groups.

among the groups, my main point is to show that multicellular organisms (the labeled groups) have arisen independently a number of times, over millions of years during the course of evolution. Thirteen separate inventions of multicellularity are indicated in the figure, but this is far below the actual number because some of those are well known to encompass multiple origins, that is, to be polyphyletic, as I shall illustrate. My descriptions will begin with animals and fungi and move down the tree in the figure.

The descriptions themselves will be brief, and to make them easier to fix in one's mind, some of the more important forays into multicellularity will be accompanied by figures.

## Eukaryotes

### *Metazoa*

We know little about the transition stages in animals. Among forms living today there is a huge gap between the sponges and their single-cell ancestors. Either all of the known invertebrates, living and fossil, came from one multicellular ancestor; or possibly there was more than one, and sponges had a separate origin.

### *Fungi*

Fungi are a heterogeneous group and the possibility that they invented multicellularity more than once is a reasonable hypothesis. All fungi are filamentous in nature, and they are to varying degrees syncytial or multinucleate, especially in the growth phases of their life cycle. One

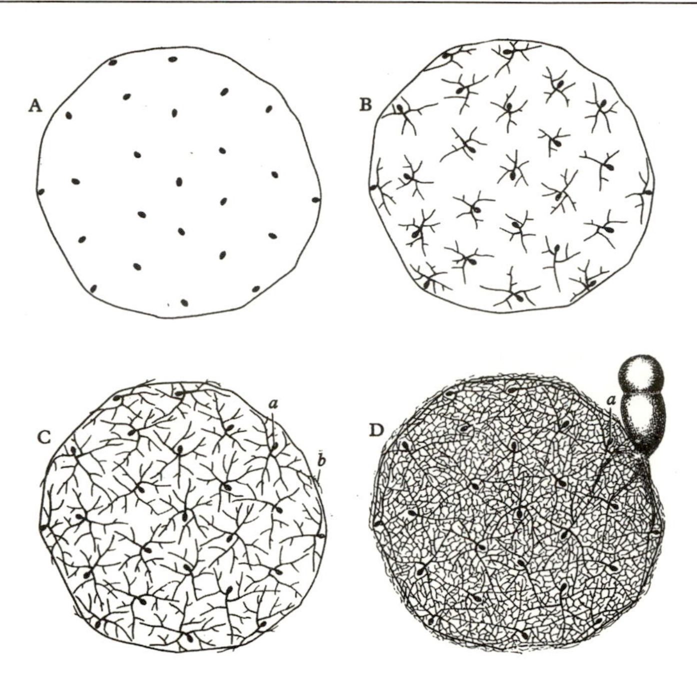


Fig. 2. The development of the small mushroom *Coprinus.* The spores germinate (A), and as the filamentous hyphae take in nutrients from the agar medium, they grow and anastamose (B, C). Ultimately, all the cytoplasm and the nuclei in the network of connected hyphae flow into a small bud to form a mushroom. (From A.R.H. Buller, *Researches on Fungi,* Longmans, Green, 1909.)

very common aspect of their multicellularity is that when they go into their reproductive, or fruiting, phase, all the nuclei and cytoplasm suddenly flow through the hyphae to central collection points and rapidly produce spore-bearing fruiting bodies. These can be quite small, as in the case of simple molds, or they can be very large mushrooms (fig. 2).

### *Green Algae and Green Plants*

The green algae began their multicellularity in water. They provide some splendid separate examples of aquatic origins in the form of what is traditionally known as colonies. It is assumed that all higher plants came from green algae in which the cells had moderately rigid walls and the division products of an asexual spore or a zygote remained glued together. This is well illustrated in the sea lettuce *Ulva* and its smaller relatives where, during the course of their development, one can follow the transition from a simple filament to a thickened thallus (fig. 3).

A somewhat different mode of becoming multicellular is seen in the Volvocales, in which the division products are surrounded and held together by jelly and the colonies they form may be flat. But more often they are hollow spheres typified by *Volvox*, the largest member of the group (fig. 4).

The Chlorococcales become multicellular in a radically different way. For example, in *Pediastrum* the products of cell division are confined within the mother cell (in a vesicle that lined the mother cell). While the daughter cells become detached and swim about using their flagella, they are initially imprisoned within the vesicle, and eventually, as they lose their ability to move, they become cemented into a flat plate that will burst free from the vesicle as they grow (fig. 5).

*Hydrodictyon* is closely related to *Pediastrum* but differs in that it produces, by much the same development, a huge colony. Finally, I should mention the coenocytic or multinucleate green algae such as *Caulerpa*, a large marine form

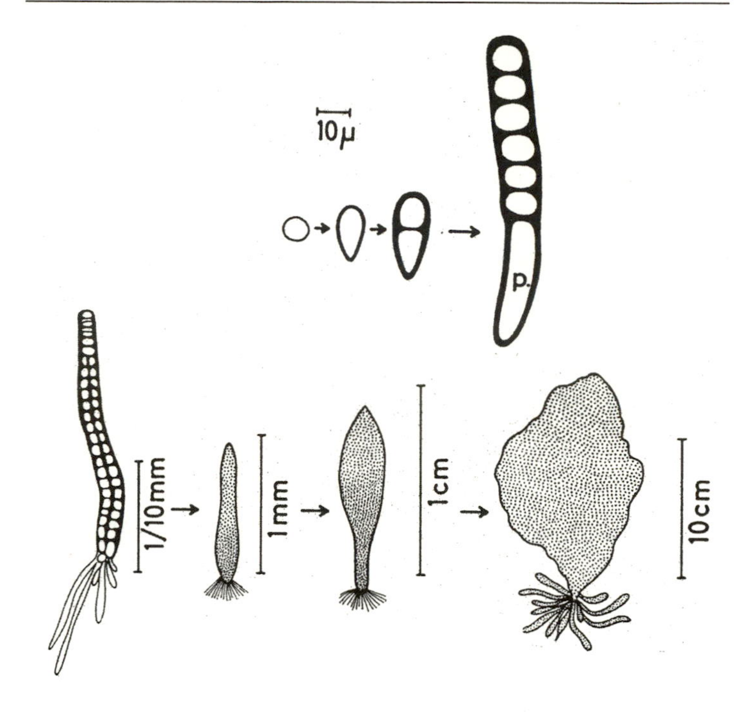


Fig. 3. The early development of the green alga *Ulva*. (From A. Lövlie, *Compt. rend. trav. lab. Carlsberg,* 34[1964]: 77–168.)

that is attached to the ocean floor by a holdfast. Inside there are no cross-walls, but merely a large vacuole surrounded by streaming cytoplasm containing vast numbers of nuclei.

The question why there is such a variety of shapes in these algae was addressed many years ago by J. R. Baker (1948), who pointed out that photosynthetic algae did not need to develop an elaborate feeding mechanism to catch particulate food; all they needed was to be able to catch the

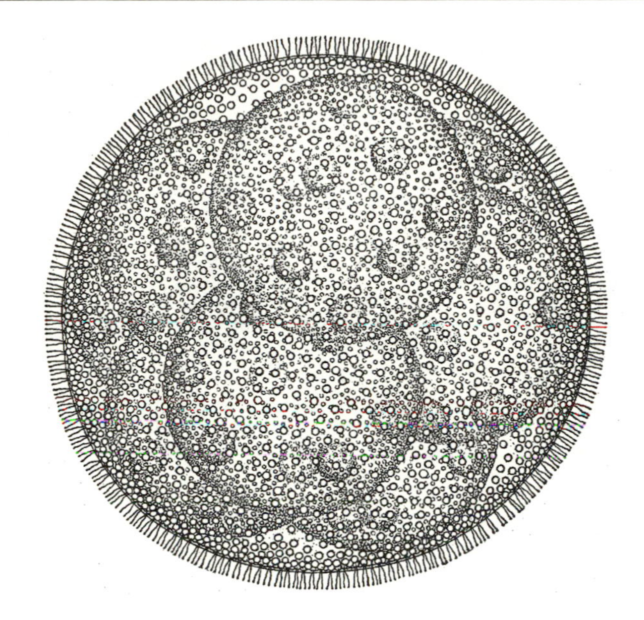

Fig. 4. *Volvox carteri.* A mother colony containing daughter and granddaughter colonies. (From W. H. Brown, *The Plant Kingdom,* Ginn, 1935.)

sun, making it possible to invent a great variety of different shapes, which is exactly what they have done.

*Brown Algae and Diatoms*

The brown algae are largely marine and are notable for having some forms, such as *Macrocystis,* which are over one hundred feet in length. It is assumed that their method of initially becoming multicellular was much the same as I have already described for the green alga *Ulva,* that is, they started as cells with rigid walls that failed to separate upon division.

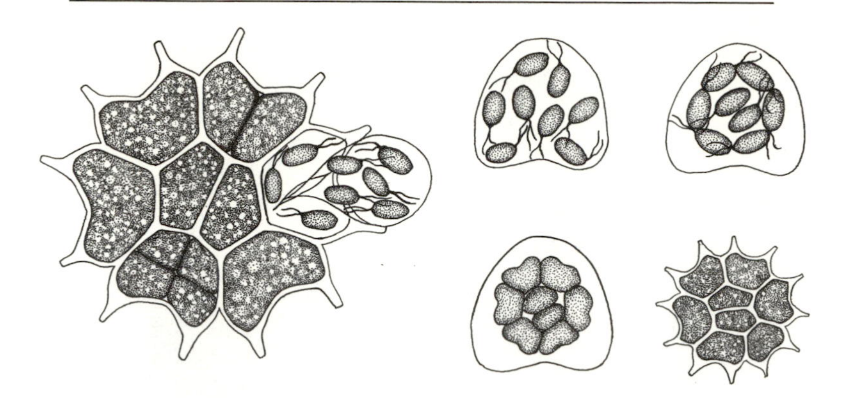

Fig. 5. Colony formation in *Pediastrum*. (*Left*) A mature colony undergoes division in some of its cells, which are then liberated in a vesicle (*right*), and the swarmers proceed to form a flat plate that becomes a new colony. (Based on drawings of G. M. Smith and J. G. Moner.)

Diatoms, which are related to the brown algae, would seem to be quintessential unicellular forms; each cell is encased in a hard silica shell. There are in fact two minor but interesting exceptions within the group. There are a few species in which the dividing cells remain attached at one end to form sessile colonies (e.g., *Licmorpha flagellata*). The other exception, species such as *Navicula Grevillei*, is particularly odd: all the motile cells secrete a tube that surrounds them and expands as they multiply. This tube, which branches, is anchored to the ocean floor. The secreted house may be a centimeter or more in height, and the separate cells actively move about inside it (fig. 6).

*Ciliates*

The cell structure of ciliates is highly specialized and different from all other organisms. Ciliates are unique in having a separation of the germ-line micronucleus and a huge

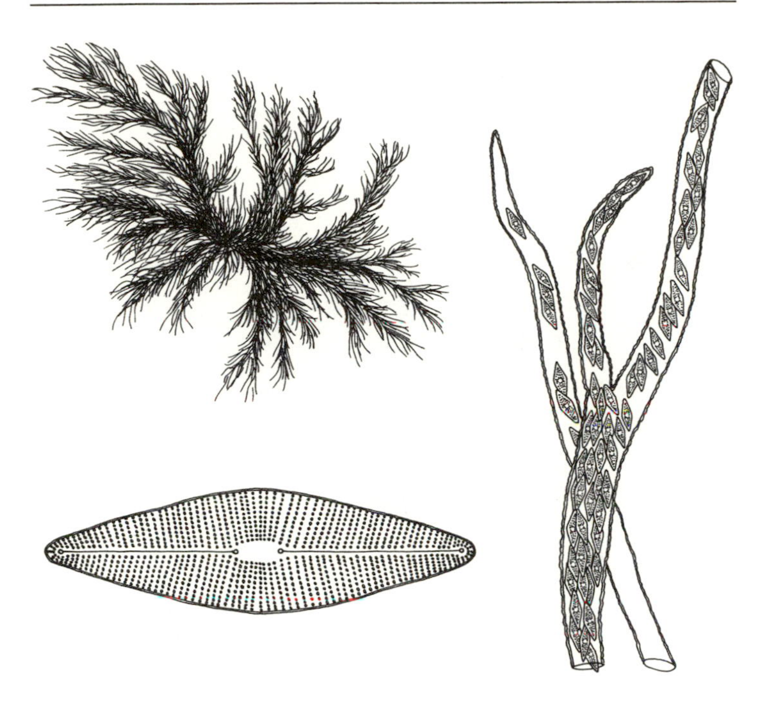

Fig. 6. (*Top, left*) A colonial diatom (*Navicula Grevillei*) showing the entire colony (about 1 cm long): (*right*) a high-power view of the diatoms wandering about inside the tube that they secrete and enlarge, and (*lower left*) a high-power view of an individual diatom. (After W. Smith, *A Synopsis of the British Diatomaceae, &c.*, London, 1856. Redrawn by Margaret La Farge.)

macronucleus that controls the morphogenetic events of the complex cortex. It could be argued that this is another way of becoming larger—one that does not require multicellularity. However, there are a number of genuine multicellular, or colonial, forms which exude a supporting adhesive thread at one end of the dividing cells, ultimately building a sessile colony of individual but connected cells.

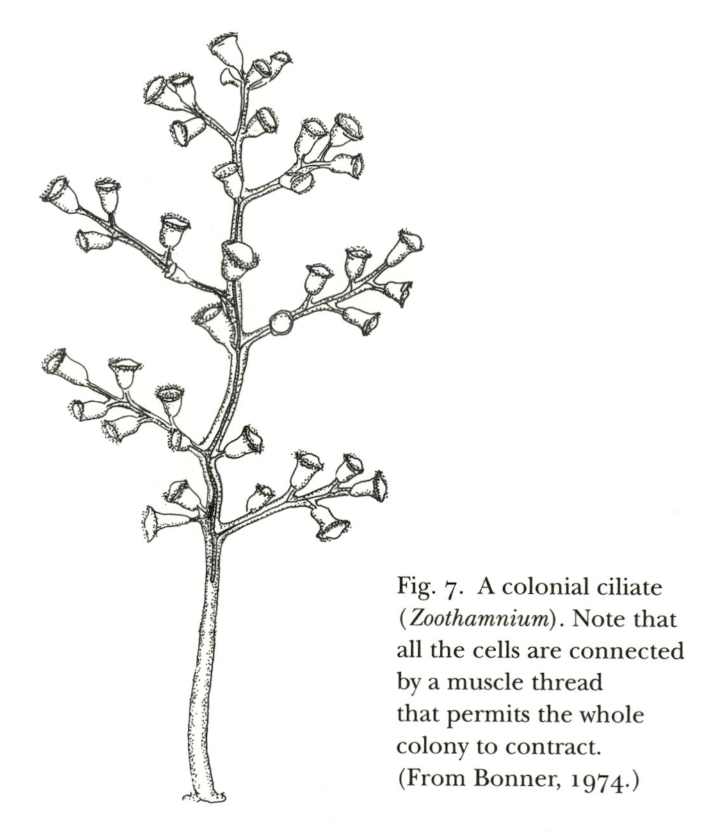

Fig. 7. A colonial ciliate (*Zoothamnium*). Note that all the cells are connected by a muscle thread that permits the whole colony to contract. (From Bonner, 1974.)

In *Zoothamnium*, the cells are also linked by a single muscle thread, so that if one cell is touched the whole colony will contract to avoid danger (fig. 7).

One particularly curious form of multicellularity is found in *Sorogena*, a ciliate that lives in the soil. When its food has been depleted, the separate cells aggregate to form a small fruiting body that sticks up into the air (fig. 8).

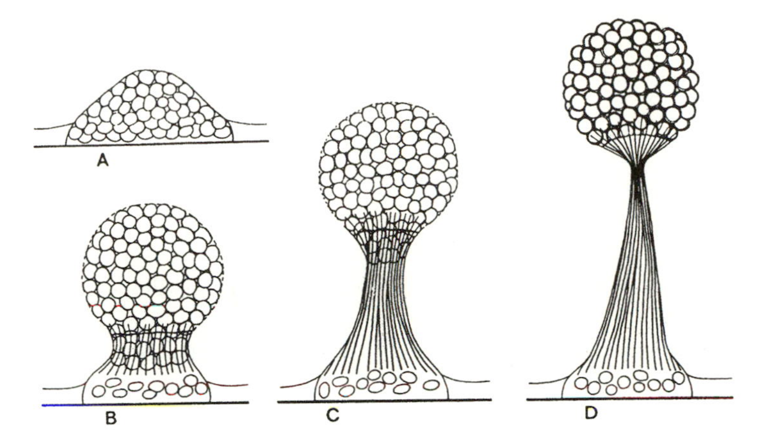


Fig. 8. The formation of the colonial ciliate *Sorogena.* The cells aggregate and rise up on a stalk they secrete. Ultimately, each ciliate cell becomes encysted. (From L. S. Olive, *Science* 202[1978]: 530.)

*Red Algae*

Although red algae have many multicellular forms that are shaped like some of the green and brown algae, their cytological and biochemical characteristics are quite distinct and set them apart. Nevertheless, it is assumed they achieved their multicellularity in the same fashion previously described for the other algae. Because they differ in so many details of their cell structure, they must have made the step to multicellularity independently.

*Cellular Slime Molds*

These organisms are characterized by an asexual life cycle in which they feed as separate amoebae. When their food supply is consumed, they aggregate into collections of cells

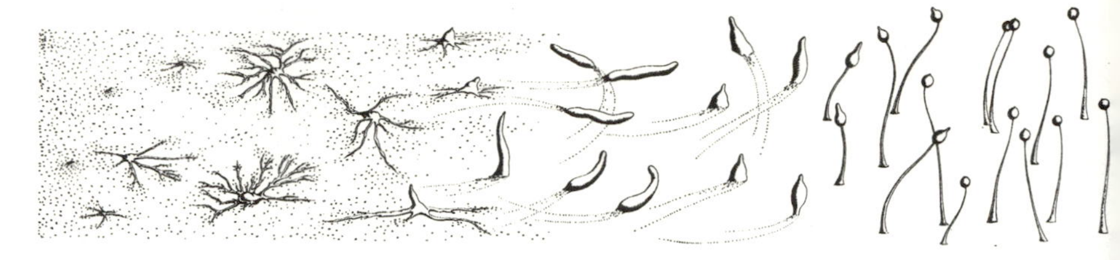

Fig. 9. The life cycle of the cellular slime mold *Dictyostelium discoideum* from the feeding stage (*left*) through aggregation, migration, and the final fruiting (*right*). (Drawing by Patricia Collins from Bonner 1969, *Scientific American.*)

that form small, multicellular fruiting bodies bearing resistant spores (fig. 9).

The cellular slime molds consist of two major groups, the Acrasids and the Dictyostelids. They clearly have separate origins for they differ in their cell structure and they are quite far removed when examined by genetic analysis.

*Myxomycetes*

True slime molds are quite different than the cellular slime molds. They have a sexual cycle in which the zygote begins as a uninucleate amoeba, but as it feeds and becomes larger only the nuclei divide and it develops into a large, multinucleate mass of naked cytoplasm—a plasmodium. When the conditions favor fruiting, the plasmodium breaks up into a single aggregate, or more often many small ones, each of which forms a fruiting body bearing haploid spores that give rise to the gametes of the next generation (fig. 10).

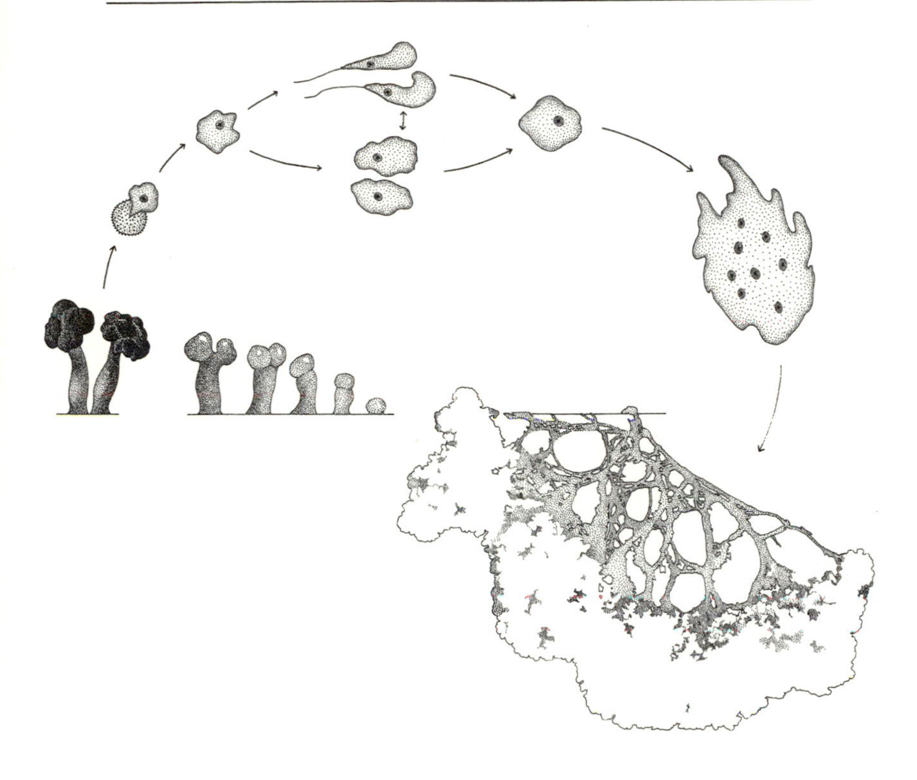

Fig. 10. The life cycle of the myxomycete *Physarum polycephalum*. The minute spore germinates (*upper left*), giving rise to a cell which, depending upon the environmental conditions, is either an amoeba (dry) or a flagellated cell (wet). After fertilization, the zygote grows into a large multinucleate plasmodium that eventually turns into many spore-bearing fruiting bodies. The lower drawings are at low magnifications; the upper ones are greatly magnified. (From Bonner 1980. Copyright © by Princeton University Press. Drawing by Margaret La Farge.)

*Foraminifera and Radiolaria*

These are amoebae with beautiful shells made with calcium or silica. They mostly float free in the ocean—a few are sessile. As they grow they become multinucleate, and in the foraminiferans they secrete additional chambers as their size increases.

### Archaebacteria

Archaebacteria are a relatively recently identified group of prokaryotes that are quite distinct in many of their biochemical characteristics from eubacteria. These ancient organisms are notable in that they can exist in many extreme environments. There are reports that some species of *Methanosarcina* form groups of compacted cells that seem to adhere closely to one another after division. It has been pointed out to me by Professor Karl Stetter that this organism is an obligate anaerobe and that these compact colonies might serve as a mechanism to keep the internal cells protected from a sudden influx of oxygen in the environment.

### Eubacteria

*Myxobacteria*

When these motile rod-shaped bacteria grow and divide, the daughter cells remain close to one another, forming a wandering swarm that steadily increases in size. When the conditions are right, these masses of rods come together in central collection points to fruit—to form cysts or micro-

Fig. 11. The life cycle of the myxobacterium *Chondromyces crocatus.* A mature fruiting body (*lower left*) bears cysts, each one of which liberates numerous motile bacterial rods, which swarm progressively into larger groups, ultimately producing a new fruiting body that rises up in the air. The fruiting body sequence is at low magnification (a mature one is about 1 mm high), while the swarming sequence is at high magnification (each bacterium is about 4 μm long). (From Bonner 1980. Copyright © by Princeton University Press. Drawing by Margaret La Farge.)

spores. In some species, such as *Chondromyces,* the cysts are lifted up into the air on a stalk (fig. 11).

*Cyanobacteria*

The cells of these eubacteria are large—as large as many eukaryotic cells. They are mostly multicellular, the usual form being linear filaments; however, there are a number of species with branching filaments. Although they are primarily aquatic, they are extraordinarily hardy and can withstand considerable exposure to air. They do form resistant spores, and some species have specially differentiated cells (heterocysts) to fix nitrogen from the air, for their nitrification enzymes can only operate in the absence of oxygen, the product of the neighboring cells' photosynthesis (fig. 12).

*Actinomycetes*

These soil bacteria form small, branching threadlike filaments. Some of those filaments will reach up into the air and produce spores. Often the spores bud off a linear series, although in some species there is a spherical mass of spores at the tip of a filament. They are a large and diverse group; perhaps the best known in *Streptomyces*, which produces the antibiotic streptomycin (fig. 13).

## A Note on Sexuality vs. Asexuality

There is an interesting point here concerning all of these different explorations into multicellularity. Unlike so many advances in evolution, sex does not seem to have been required for the invention of multicellularity. Clearly, sexuality exists among unicellular organisms; in other words, sex undoubtedly antedated multicellularity. There is no obvious correlation between inventing multicellularity and sexuality. Primitive multicellular organisms may

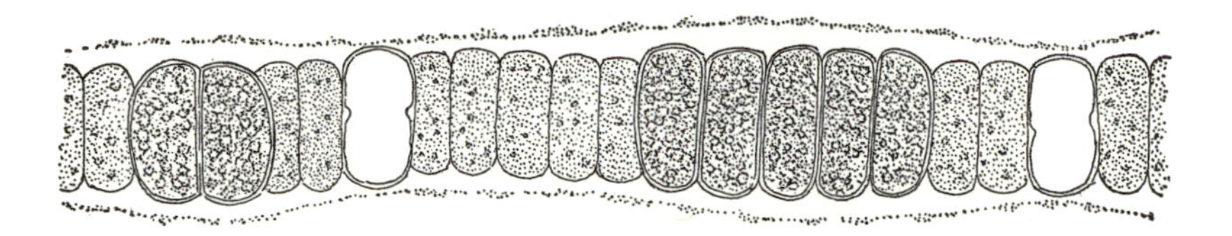

Fig. 12. A filament of the cyanobacterium *Nodularia* showing vegetative cells, the chlorophyll-free heterocysts where nitrogen fixation take place, and some thick-walled spores, or akinetes. (From G. M. Smith, *The Freshwater Algae of the United States*, McGraw-Hill, 1933.)

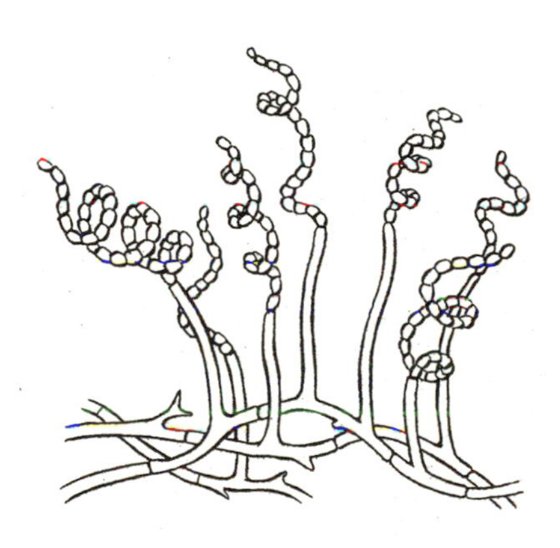

Fig. 13. The aerial hyphae of the actinomycete *Streptomyces*. (Drawing by Hannah Bonner.)

either produce asexual spores or gametes. Among the fungi and the green algae one finds species in which there is an alternation of sexual and asexual cycles, both of which will produce spores or some form of unicellular propagules.

## The Variety of the Mechanical Ways of Becoming Multicellular

If we now look at these various experiments in inventing multicellularity, we want to see if some general statements about their different mechanisms can be made. We are looking for a macro-explanation, for a holistic view. One can make a sharp distinction between those organisms that became multicellular under water and those that did so on land. All the aquatic organisms began their multicellularity when the products of cell division failed to separate, while most terrestrial microorganisms involve some form of motile aggregation of cells or nuclei in a multinucleate syncytium. The separate origins of terrestrial organisms are indicated by the open rectangles in figure 1.

### Aquatic Origins

In the case of the aquatic origins there is a wide variation, and this variation seems to be correlated largely with the type of cell, and especially the type of cell surface. For those organisms with cell walls, one of the most prevalent methods of building a multicellular body is to have the cells fail to separate after division. The hard polysaccharides of the cell walls found in green, brown, and red algae have permitted the rise of filaments and ultimately the branching and thickening of those filaments to form solid tissues and produce large plants. The same mechanism in its filamentous form is found in the cyanobacteria and some other eubacteria. All of these forms use their cells as building blocks once they fail to separate.

Animal cells do not have a cell wall, and they stick together because of adhesion molecules on their surface membranes. That they are covered only by a membrane is no doubt related to the fact that originally the cells were amoebae and they had to retain a pliable membrane so they could engulf bacteria and other food particles. Consistent with J. R. Baker's point mentioned earlier, the next big mechanical step must have been devising a feeding mechanism for a group of cells, but any clue as to how this was achieved initially has been long lost.

Diatoms and ciliates have highly specialized outer coverings, which present another kind of problem. In diatoms it is the rigid silica shell, while in ciliates it is the elaborate cortex with its cilia and other complex structures—in both cases an external armor raises a challenge to cell adhesion. Yet a few species of both groups have found an identical solution. By exuding an adhesive stalk at one end of the elaborate cells, they can divide without interference from the adhesive, and the daughter cells will be attached to one another at one end—by their roots, so to speak—and in this way form a branching colony.

Then there is the genuinely curious solution, also previously described, of the diatom that surrounds itself with a branched, secreted tube as it multiplies. It is as though it were creating a miniature biosphere.

## Terrestrial Origins

In microorganisms, it is difficult to draw a sharp line between aquatic and terrestrial, because the latter need water too and will always exist in a thin layer of water covering particles of soil or humus. In other words, all terrestrial microbes are to some degree aquatic. This is especially true of their feeding phase, for almost none of the multicellular terrestrial microorganisms are photosynthetic, and all require a liquid film. If they are particle feeders, they need this to move about and engulf their food. If they are saprophytes, they must get the food itself in liquid form.

These terrestrial organisms send a spore-bearing body that pushes up through the water interface into the air for dispersal. This is certainly true for the fungi with their enormous array of aerial fruiting bodies (although the water molds have similar sporangia that form and liberate motile zoospores below the water surface), and it is also the case for both kinds of cellular slime molds as well as the true slime molds, or myxomycetes. Aerial spore-bearing bodies are also characteristic of terrestrial multicellular prokaryotes: some myxobacteria produce fruiting bodies with prominent stalks while others merely form mounds, and the actinomycetes have chains of spores sticking up into the air.

The mechanics of aggregation, which is characteristic of most terrestrial microorganisms, takes two forms: it involves either the gathering of separate uninucleate cells, or the gathering of nuclei and cytoplasm in a multinucleate syncytium.

In the case of aggregating single cells, we have a remarkable bit of convergence, for aggregation occurs independently in eubacteria, in two distinct kinds of cellular slime molds, and in ciliates; in other words, it has arisen independently at least four times. And there are numerous variations in the mechanics of fruiting-body formation within some of the groups. In the cellular slime molds, for example, the fruiting body can be produced (1) by the cells squirming past one another, whereupon the terminal cells that have been pushed to the top of the stalk they have formed become spores; (2) by the cells making a stalk of dead cells by adding cells at the tip, in the process lifting the cells that become spores; (3) by secreting a slender, cell-free hair of cellulose, causing the cells to rise upwards and turn into spores; (4) by secreting a noncellular cone and using osmotic pressure to push the spores up to its narrow end, like an erupting volcano. This latter mechanism is interesting because it is similar to the method used by the totally unrelated ciliate, *Sorogena*; again this is a striking case of convergence (see fig. 8).

In those instances where there is a multinucleate syncytium, the aggregation stage is different; it involves the directed movement of a great mass of protoplasm toward central locations that are the incipient fruiting bodies, as we saw in fungi and myxomycetes. The multinucleate mass in myxomycetes, which can be very large, is covered only by a thin membrane and therefore requires very moist conditions on the forest floor or on dead logs. This plasmodium is the feeding stage, which is separate in time from the fruiting or dispersal stage.

The very same separation of the feeding and fruiting phases is found among the great multitude of species of fungi. However, there is one big difference. The syncytium is always confined in a chitinous tube, the hypha. As these hyphae branch and become a mycelium they will invade the soil or rotten wood—whatever might be their substrate—sopping up nutrients and continuously expanding and enlarging. When the right conditions arise, all growth ceases, and the protoplasm surges back through the hyphae to a central collection point (see fig. 2). It may be a modest surge and produce a simple spore-bearing mass held up in the air by a single hypha, as in *Mucor*, a bread mold, or it may involve the aggregation of a large amount of protoplasm in a vast mycelium to form a mushroom (which can be very large indeed, as in a giant puffball that weighs many pounds).

## When Multicellularity Occurred in Earth History

There is a general point concerning the origins of multicellularity that needs emphasis. The moment in evolutionary history for each of the different organisms described here occurred not only as separate events but at different times. There is good evidence that multicellularity in cyanobacteria was invented at a very early stage, a good 3.5 billion years ago (Schopf, 1993), yet the first multicellular animal was probably a much more recent event. Unfortunately, we do not know the sequence in time of the origins of the various groups of organisms shown in figure 1, but they must span an eon. Indeed, there is nothing to rule out the possibility that at this very moment multicellularity is in the process of being invented by some single-cell form somewhere on our earth.

## What Might Be the Selection Pressures That Produced Multicellularity?

### *Aquatic Origins*

The fact that multicellularity arose independently so many times is the primary basis for believing that there has been a significant selection for it in the ancient unicellular world. Yet it is difficult to guess what the first advantages might have been. In the case of aquatic origins, it is easy to imagine that the failure of the daughter cells of division to separate might be the result of a simple mutation. It is the next step that is harder to picture: what advantages would clusters of cells have over single cells? Perhaps initially they had neither advantage nor disadvantage and survived by drift until some further mutational change endowed them with a skill that was not possible for their single-cell relatives. I will pursue this possibility presently; here I am only concerned with the initial step.

It might be that the mutation that allowed the cells to adhere to one another also allowed them to stick to the substratum. Under some circumstances, where the cells in an ideal location for growth are likely to be swept away by currents, remaining fixed to one spot might be selectively advantageous. The same advantages would apply to the daughter cells, thereby giving rise to a multicellular sessile colony. Such an occurrence might be the origin of colonial stalked ciliates and diatoms. This could even apply to the diatom that is encased in a tube—instead of the adhesive that glues the cells to the substratum, a different set of mutations allowed them to build a cocoon of stiff material around them inside of which they could divide and grow. In these instances, becoming multicellular is the inevitable consequence of the advantage of remaining in one spot.

I have said very little about foraminiferans and radiolarians. They are mostly free in the ocean, or pelagic, and as they grow they become multinucleate. But what is the advantage in their doing so? One argument has been put forth by G. Bell (1985) for *Volvox* that might apply here: the size increase prevents filter feeders from being able to eat them. The problem for this ingenious argument is that we are looking for the origins, and the "too big to eat" idea requires that there are already large multicellular predators about. Perhaps initially, pelagic multicellularity had no advantage—organisms just grew.

In the case of primitive flagellated or ciliated cells, it is conceivable that size increase is an advantage, because the larger the organism the faster it will swim. Again, it is easy to see that in an aquatic world where there are a multitude of different-sized organisms, being large and fast helps to catch prey (or to escape), but again that does not help us to explain the first step. There lurks the possibility that initially there was no advantage, and it was only later that it was retained because of positive selection pressure.

## Terrestrial Origins

The incredibly large number of different organisms existing today that have fruiting bodies suggests that there

Fig. 14. The protostelid *Nematostelium.* A single amoeba secretes a stalk, lifting itself up into the air and turns into a spore. (From Olive and Stoianovitch 1966.)

has been and still is an enormously strong selection pressure for the dispersal of spores, cysts, and even seeds in higher plants. At first glance it would seem that, as before, this development of a fruiting body must have been something that occurred well after the appearance of multicellularity, but let us examine the possibilities a bit more carefully.

In the case of cellular slime molds there is a unicellular relative, *Protostelium,* that makes its own stalk so that it rises up into the air to form a spore (fig. 14).

Clearly, if a fruiting body could be made with numerous cells it might be even more effective in dispersal. For a soil amoeba, where feeding must be done as single cells by phagocytosis, the aggregation of cells is required for achieving multicellularity. The question of how this arose is more difficult, but a species of the distantly related soil amoeba *Hartmanella* forms resistant cysts in clusters (Ray and Hayes, 1954). There is an aggregation of the amoebae before encystment, yet no fruiting body. We do not know enough about dispersal mechanisms in the soil and can only ask whether this primitive clustering of cysts might somehow enhance dispersal.

There is an interesting case in the cellular slime molds where Kessin and his co-workers (1996) showed that in their larger, multicellular state they were unable to be eaten by nematodes, their main predator—the worms could not penetrate the collective slime sheath which protected the amoebae inside.

In the case of syncytial forms the multicellularity is more closely associated with feeding. In both the myxomycetes (a eukaryote) and the myxobacteria (a prokaryote), it is clear that by an increase in size of the feeding mass they can feed more effectively. They produce extracellular enzymes that digest large particulate food, which they then absorb directly. In myxobacteria Dworkin (1972) has called this "wolf-pack feeding." So in both these cases we could guess that multicellularity arose as an advantage in feeding, and the formation of fruiting bodies was secondarily derived because of the advantage of effective dispersal. The same arguments would apply to the fungi.

## Conclusions

What we see in this examination of the large array of experiments in multicellularity is that in early evolution becoming larger took on many forms; in fact, no doubt there are others we do not even know about which have already gone extinct. As I have argued, the most reasonable guess is that originally they arose by chance mutation and subsequently were selected because of some advantage that they might accidentally have accrued. This initial success was often greatly improved upon, as we saw for cyanobacteria and cellular slime molds; but the raw materials for natural selection had been laid down by cells accidentally clumping together.

These days we have become beguiled by diversity: how animals, such as insects, and plants, such as angiosperms, have produced so incredibly many species. In the origins of multicellularity we see a most primitive example of diversification. In some ways it is almost an ideal case, because we can make an argument for its basis: size increase is the common cause of all the small successes that I have described.

And it is this very cellular diversification that will be so interesting to examine in our future inquiries. What are the biochemical differences and similarities among the adhesives that are used in the various aquatic forms that invented multicellularity? Is there a genetic connection between the separate ones, or are some or all of them unique? The same questions can be asked about chemotaxis in terrestrial forms: what are the molecular connections, if any, between the different chemoattractants? And adhesion plays a role in aggregation too. Finally, there is an especially fascinating question. In the first signs of cell differentiation, there is both a mechanism to alter the fate of an individual cell and a mechanism to place the differentiated cells in a regulated, controlled pattern. What are the similarities and differences in the molecular mechanisms of all the independent inventions of these remarkable phenomena? I think the study of the origins of multicellularity has a bright future.

# 4 *Size and Evolution*

THE MOST OBVIOUS reason we have multicellular development lies in natural selection. If we could see and understand the external influences that caused development to arise and evolve, we would grasp the origins of development and win some insight into its primordial mechanics.

Selection works in two cardinal ways on the life cycles of all organisms. There is a selection pressure for size increase under innumerable ecological circumstances, and becoming multicellular is an easy way to accomplish this. At the same time, as I have argued earlier (1958), there is a selection to retain a single-cell stage in the life cycle. This is a requirement for sexual reproduction, for meiosis and fertilization can only be achieved in a unicellular stage in eukaryotic organisms. The reason has to do with the way the genetic material is incorporated into chromosomes, and with the separation (and recombination) of the allelic pairs of genes at meiosis. Furthermore, the fusion of the genomes of two parents can only take place in single cells. A unicellular stage is also favored in many asexual organisms where dispersal in the form of spores is best accomplished by the smallest bodies that are easily carried to some new feeding ground.

This leads us to a paradox of cosmic implications: natural selection is simultaneously pushing for a large stage in the life cycle that can compete for food and for a minute single-cell stage that is essential for sexual reproduction—and often asexual as well. The result is that all multicellular organisms, from small algae and fungi to elephants and giant sequoias, have a unicellular stage and a large stage of varying dimensions in their life cycle.

Furthermore, because of this schizophrenic simultaneous selection for the large and the small in a single life cycle, there is a corresponding separation in time between the feeding and the reproductive stages of the cycle. It is a general principle that the stage of the life cycle that involves the greater intake of energy is the larger multicellular stage. In plants, larger body size means more surface for photosynthesis; in animals, it means more efficient capturing of prey or resisting the attack of predators. Protection, as a general phenomenon, is linked to feeding. There are some apparent exceptions, such as in the cellular slime molds, where feeding occurs at the unicellular stage, and the spore-bearing reproductive body is multicellular. The single-cell feeding stage is indeed a genuine exception to the rule, but the large fruiting body has arisen to disperse the unicellular spores more effectively. It could be argued that dispersal to new patches of bacterial food is a part of the energy-intake mechanism, although precise definitions of the stages of these unusual life cycles is perhaps an empty exercise compared to the interest of their exceptional nature and how they adapt to their particular environmental circumstances. The life cycles of the vast majority of organisms fit the rule of a large feeding stage alternating with a minute reproductive stage.

## Selection for Size

I am assuming that the selection to keep the unicellular stage in the life cycle is invariant, and therefore it will not be of concern here. On the other hand, the size of the feeding stage is of primary interest because it varies so much in different organisms that one wants to know why sometimes the size increases and sometimes it decreases during the course of evolution.

It has been my conviction—a conviction that has been growing with the years—that size selection is of paramount importance in evolution. There are many reasons why it has been much neglected in the literature: it has taken on a bad name because it is sometimes associated with the uncomfortable notion of "progress." Even more significant is the fact that the form, the shape of an organism, is intrinsically more interesting than its size. I want to discuss these points in more detail to support my argument that selection for size is central.

It is a fact that there has been an increase in the size of organisms since the beginnings of life, when the largest organisms were bacteria, to modern times, when we have huge beasts and trees. Furthermore, this increase has been gradual over time, as many, myself included, have argued before (1965, 1988). This trend is a simple reflection of the fact that there is always an open niche at the top of the size spectrum; it is the one realm that is ever available to escape competition. Under a variety of ecological conditions it is equally possible to evolve smaller as well as larger organisms depending on what niches are open. This means that the overall size trend can be explained in terms of the expanding upper limit of the size niches and natural selection; there is no need to invoke any mysterious intrinsic trend such as the term "progress" implies.

Much has been written to show that for any group of animals or plants—or even for protists (such as radiolarians) that have left a fossil record—with time their size sometimes increases and sometimes the trend is a decrease in size. Size increase was more often emphasized; for instance, it even has a name, "Cope's rule," while size decrease bears no such honor. There were, especially among vertebrates, many beautiful examples of size increase over geological time, such as the evolution of the camel, and of the horse from the early dog-size *Eohippus* to the large *Equus* of today. These trends, freshly discovered in the latter part of the nineteenth century, formed the foundation for the mystical and popular notion of "orthogenesis," where it was gratuitously assumed that there was some inner force that pushed evolution towards bigger and bigger things. It was only much later, well into the twentieth century, that the paleontologists showed that downward trends, although perhaps less conspicuous, were abundant as well. The only real difference was that while there is a lower limit for small size, the upper ceiling for size proved to be continuously expandable over earth history. This is the reason for the ever-increasing upper end of the size scale for animals and plants. It is not an innate, orthogenic trend, but the simple matter that all organisms evolved from small unicellular forms, and therefore there is always room at the top, something continuously exploited by natural selection.

When we think of organisms, we think of their form, of the structure that characterizes them. If we see a small tree and a large mammal we only note in passing that one is bigger than the other—that seems an insignificant and obvious fact. Rather, what strikes us as much more significant is the myriad of interesting facts about their shape, the way they move, and their internal structure. We classify organisms on the basis of morphology and we study their physiology on this basis. What kind of a Linnean taxonomy would be possible if all descriptions were based on size alone—the thought is simply ludicrous.

Moreover, we know that natural selection can and does act directly on the form and function of all organisms. No doubt there is competition among organisms that differ simply in size, but even more obvious are differences in coloration, in the swiftness of movement, in the ability to avoid losing moisture in desert plants and animals, in the ability of diving mammals and birds to hold their breath under water—and this list could be extended infinitely.

All this being self-evident, how is it possible for me to claim that selection for size is so important, yet so neglected? My argument is based on the notion of niches; size niches are one of the prime movers in all of evolution, and one of the prime movers in any ecological environment; they are the frame upon which Charles Darwin's famous tangled bank rests. They are the great lattice holding together all of life on earth.

## SIZE NICHES

Ecological communities are extraordinarily complex, and the concept of niche has been helpful in understanding some aspects of that complexity. The notion that there is a space in nature—such as a link in the food chain—that is filled by a competitively successful species has an appealing logic to it. It is something we all know is there and which we accept as the backbone of an ecological community. As a concept it is convincing and useful. It helps us to see the basis of convergent evolution. The marsupial mammals of Australia independently mimic the placental mammals from elsewhere. This fact can comfortably be accounted for by thinking that the niches in the two parts of the world are much the same and have molded species to fit them, even when the starting material is quite different. A striking example is in William Bates's old book on the Amazon (1863), where he shows that a hummingbird and a hawk moth have a close resemblance because they compete for the same food, the same niche (fig. 15).

Despite the agreeable commonsensical aspect of the concept, a niche has been difficult to define in any easy way. The problem is that there are so many facets or properties to a niche that any kind of a comprehensive definition almost immediately disappears in a fog, yet we all intuitively know what we mean by the word. I do not intend to dismiss the important contributions of many ecologists; they made it possible to see and even attempt to define the complexity intrinsic to the notion that so aptly reflects the complexity of a community itself. However, the formulations have been unwieldy.

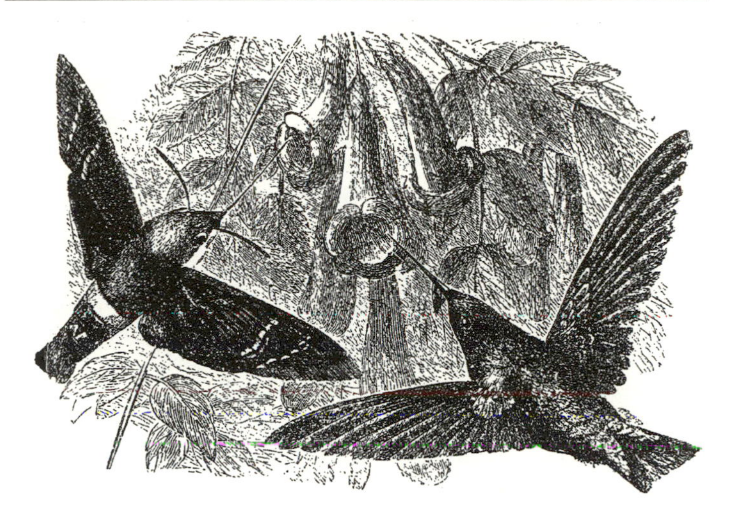

Fig. 15. A hummingbird and a hummingbird hawk moth. (From H. W. Bates, *The Naturalist on the River Amazons*, Murray, 1863.)

My intention here is to do something quite different and consider only a one-dimensional niche—the size niche. I should immediately add that G. E. Hutchinson (1959) led the way in using size niches very effectively as an important ecological concept. My justification for putting unconventional emphasis on size here is not just to have a simpler and more tractable use of the concept of niche—although that certainly is part of it—but more importantly because size in itself is a central, ubiquitous component in all of evolution. It is the one feature that all organisms have in common; almost all other traits are local. Flowering plant, algae, fungi, invertebrates, vertebrates, one and all, come in different sizes and are constantly subjected to selective forces that encourage their progeny to be bigger, or smaller, or remain the same size. Size is not only a component of all organisms, but of all environments. Selection for size is universal. Let me give an example in some detail that will help to make my point.

### Volvox *and Its Relatives*

In many ways, the volvocales provide an almost perfect illustration. They are organisms with which we are all familiar, usually beginning with our first biology course. They range from their single-cell *Chlamydomonas*—the ancestral type—to the large *Volvox* made up of many thousands of cells, and there is a beautiful series of genera with intermediate-size colonies (fig. 16). Furthermore, in *Volvox* itself large size is accompanied by a division of labor: most of the cells remain vegetative and concern themselves solely with photosynthesis and locomotion, while a few cells are either asexual or sexual reproductive cells and are able to start the next generation. In the smaller species, all the cells manage both functions.

For these reasons the volvocales size series has been considered by all to represent a simple evolutionary sequence of size increase, ultimately accompanied by a division of labor in *Volvox* itself. This has certainly been my assumption of many years' standing, only to be shattered by the work of David Kirk and his collaborators (admirably summarized along with all other aspects of the biology of the volvocales in his recent book, 1998).

Using modern methods of molecular systematics, it has been possible to show that on the basis of genetic differences there is no simple evolutionary size sequence, but rather that some small species have large species as their

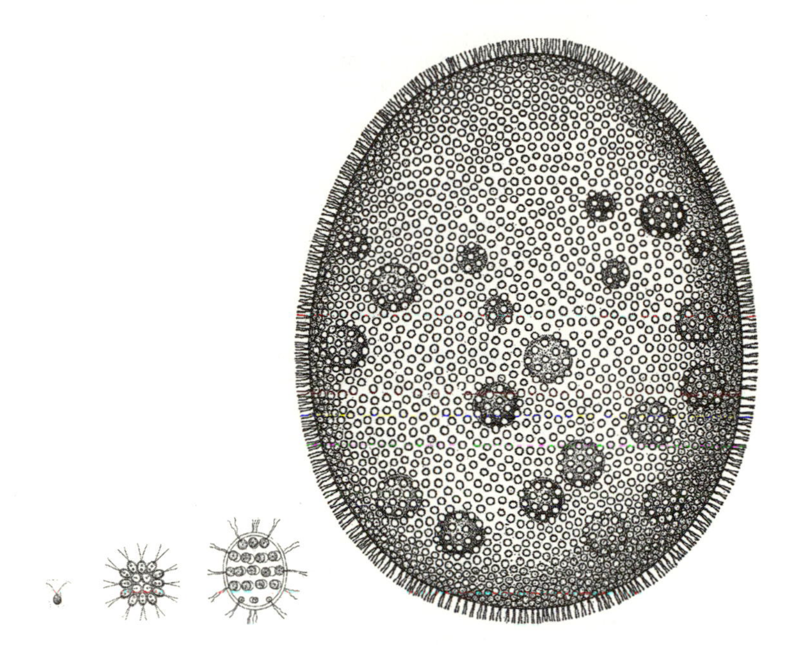

Fig. 16. *Volvox* and its relatives showing their relative sizes. *Left to right*: *Chlamydomonas* is a single-cell flagellate; next is *Gonium*, a sixteen-cell colony; followed by *Pleodorina* with thirty-two cells, and finally *Volvox* with thousands of cells. (Selected from W. H. Brown, *The Plant Kingdom*, Ginn, 1935.)

ancestors, as well as the reverse (fig. 17). The genera were not defined as genetically homogeneous groups, but rather were based solely on morphology. *Volvox* itself clearly evolved independently more than once. This means that the actual evolutionary sequence is not the same as the simple size sequence. Instead, colony size must have gone up and down with considerable frequency, although today there exists a complete size sequence, and presumably that must have been true at any one time

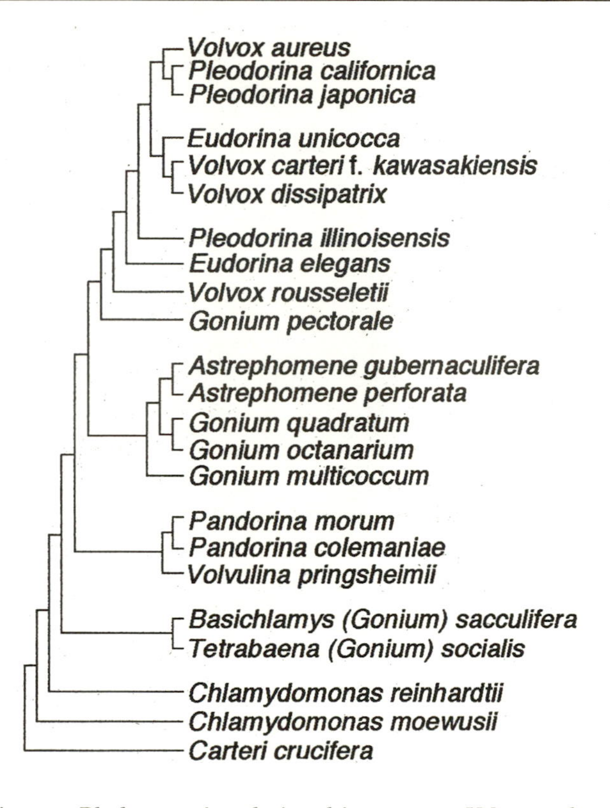


Fig. 17. Phylogenetic relationships among *Volvox* and its relatives inferred from sequences encoding for one of their chloroplast genes. Line lengths are arbitrary and are meant only to indicate the inferred branching order. Note that except for the ancestral single cell *Carteri* and *Chlamydomonas,* there is no phylogenetic size trend, and the large *Volvox* species (which have arisen independently in four separate lineages) are ancestors to small colonies. To give some idea of the size differences, *Gonium* and *Volvulina* consist of 4–16 cells; *Pandorina,* 8–32; *Eudorina,* 16–64; *Pleodorina* and *Astrephomene,* 32–128; *Volvox,* 500–60,000. (From Kirk, 1998.)

throughout their long history, although the actors were continually changing places. With such short generations there must have been extraordinarily many opportunities for size change, both up and down, over a time span of at least a billion years.

My argument from these facts is that it is the size niches that have remained constant over all this time and still exist today; what has changed are the occupants of those niches. What is it, then, that is so special about those niches? This has been examined in the most interesting way by Graham Bell and his co-workers (for references, see Kirk, 1998), and here are their main conclusions. Size increase in the volvocales produces the advantage of escaping predators such as rotifers that are filter feeders: the multicellular colonies are simply too big for such predators to sweep them into their gullets. Furthermore, the larger the colony, the faster it can swim and escape predators. Also, Bell and his colleagues have evidence that larger members of the *Volvox* series can more effectively store inorganic nutrients such as phosphorus, and this is related to their giving their asexual offspring a better nutrient start in early life. On the other hand, *Chlamydomonas* and the smaller colonial species thrive too; so, depending on the physical and nutrient conditions of the pond, the ideal environment for small volvocines must exist as well.

In fact, one of the most striking aspects of pond ecology is its complexity. There is variation in the basic nutrient levels of different ponds, each of which is affected differently by the passing seasons. This is not just true for the pond as a whole, but conditions will also vary at different depths. So, from the point of view of true Hutchinsonian niches, there is a complex and ever-changing matrix of environmental conditions involving the surrounding chemical composition of the water (which varies with the depth), the temperature and the amount of light (seasonal and diurnal), and, of special importance, the kinds of other living creatures that are part of the environment at any one moment. This is an aquatic version of Darwin's tangled bank and a true reflection of the extraordinary diversity even in the smallest spaces on the surface of the earth.

Such a picture is intimidating because of the large number of facets it embodies at one time. What I would like to suggest is a way to simplify that intricate picture and look at the world from one point of view: from the point of view of size. Size can be used as the ruler against which all else is measured. One advantage to such a scheme is that size is a universal property of all organisms. Another is that size niches remain constant over geological time, while the organisms that occupy those niches may frequently change over the course of evolution.

## Conclusion

There are a number of reasons for this emphasis on size, and let me conclude by making clear its relation to my forthcoming arguments concerning the origin of development. First, the size increase produced by multicellularity has occurred because of the selective advantage of size increase, and a number of examples have been given to illustrate this point. Second, as the upper extreme of organism sizes expanded because of this selection, the environment became structured into a series of size niches, all of which are filled with the appropriate size organisms, and the niches can be filled by either a size increase or a size decrease. In this way we have answered the question of why multicellularity has arisen in evolution, and why those first multicellular forms became larger through further selection.

In the pages that follow we will try to understand the first mechanisms of how a primeval collection of cells began some sort of coordinated development. How did cell-cell signaling arise, and how was that signaling put to use to get an integrated development, one that was consistent from one generation to the next?

# 5 *The Evolution of Signaling*

WE OFTEN think of the origin of life as fundamentally a problem of the origin of template replication. That certainly is a central property of all living things, and there is no way life could have evolved without it. It has been elegantly pointed out by Dyson (1999) that replication of itself is not sufficient and that metabolism is another important property that was essential for life right from the beginning. Here I would like to add a third element that must have been equally important in distinguishing the living from the inert. It is the invention of a stimulus-response system. Ultimately, this system became the basis of how organisms respond to the environment, and how parts of an organism and, at the highest level, whole organisms communicate with one another.

That a stimulus-response mechanism was something quite fundamental struck me in the early days for a straightforward reason (1958). In thinking about development, I saw how many parallels there were between the steps of development and the steps involved in a particular event in animal behavior. In developing animals and plants, each morphogen is a stimulus that produces a specific response, or sometimes a general response, which in turn leads to the production of another morphogen that produces another stimulus. In the case of behavior, there are innumerable examples, especially among veterbrates: for instance, in courtship an individual will make a prescribed move, and if it is the appropriate one, the partner will make an answering move, and this stimulus-response cascade will ultimately result in mating if all the reciprocal cues are right. This process is elegantly illustrated in the early work of the ethologists on courting behavior in stickleback fish, where the male and the female go through a series of motions leading up to mating, in which, in a series, each signal elicits a countersignal. Other good examples (although less elaborate ones) may be found in the exchanges between parent and offspring when feeding or in responding to moments of danger.

The question of the origin of a stimulus-response system took hold as a result of a conversation I had with my friend Jonathan Weiner. His interest was in the genetics of behavior, and he asked me about the behaviors of slime molds and other simple organisms. Orientation to light and to other external stimuli are commonplace and have been the object of much study. We wandered on to the somewhat archaic word "irritability," which was used by Thomas Henry Huxley and others in the nineteenth century to define one of the basic properties of life. Reproduction and motility were joined with irritability to distinguish the living from the nonliving.

Largely because of the great successes of biochemistry and molecular biology, the origin of reproduction by replication has not only taken center stage, but pushed all other considerations off the stage. The origin of motility is a physico-molecular problem of considerable interest, and as we learn more about the most basic contractile proteins, this subject is receiving increasing attention. However, there has been little or no interest in how irritability might have arisen, and it is one of the central properties of all living things.

## Beginning Stimuli

It is obvious that the very first stimulus to a presumed primordial replicating molecule, such as RNA, must have been from the immediate environment. Perhaps it began simply as the presence or absence of conditions which allowed or prevented replication. If in some particular RNA molecule there was a base substitution that allowed for a more rapid response to the external changes, then that particular modified molecule could have had a selective advantage and flourished at the expense of the slightly less responsive ones. Any such considerations are so highly speculative that they are hardly satisfactory; rather, one must imagine that there are many ways that outside cues could affect replication, and that this could become an opening wedge for natural selection to begin its work.

A harder step for realistic speculation is how a consistent stimulation could have been produced by the replicating system itself. One obvious possibility is that, again through some base change, some degree of autocatalysis arises. Once replication starts, it stimulates itself to go faster. The next step might be that it also receives cues from other neighboring molecules, which under some circumstances conceivably might stimulate replication to its selective advantage. This fits in with the idea that strings of RNAs have evolved enzymatic properties, making them an excellent candidate for the source of the first primordial stimulus.

However crude and elementary these stimulus-response beginnings might have been, if natural selection is allowed to intervene they can evolve into something more dependable and, ultimately, more sophisticated. It is perhaps pointless to try to build an elaborate hypothetical scheme of how the origin of a stimulus-response system might have come into being; what is important is to realize that something of this sort had to have happened, and that it is the foundation of what ultimately evolved into development and behavior. And even more obviously it is the basis of all of physiology of all organisms, from the lowliest prokaryotes to higher plants and animals. Those early beginnings spawned taxes and tropisms that have led ultimately to complex human behavior.

As we go up the evolutionary ladder of size increase, far removed from the very beginnings of the origin of life itself, one place we can look for the early uses of the stimulus-response system is in the origin of multicellularity. One might ask, Why not begin with the beginning of the cell, which after all is the next big step? I have two answers to that question. One is that we would have to compound the degree of speculation manyfold; in fact, to such a degree that for me it would be totally unsatisfying. The time may come when this is possible, but I do not think we are there yet.

The other reason is that multicellularity allows us to have the opportunity to see the birth of a new set of signals and responses that occur between cells—they are now, so to speak, out in the open and not all shut up within the cell. It provides us with a chance to look for simple, basic explanations for the process of development. Furthermore, we have many different kinds of organisms that exist today that are just over the brink of multicellularity, so we can deal with concrete examples.

## The Evolution of Morphogens

If one surveys the various signaling molecules, or morphogens, during the development of different organisms, one finds a great range in their sizes, from volatile ones such as ammonia, nitric oxide, oxygen, and carbon dioxide, to medium-size molecules that can diffuse readily through water, and finally to large proteins that are too big for rapid diffusion and must be passed from one cell to another when the cells are touching. It would take a few seconds for a volatile substance such as oxygen to diffuse a millimeter, a molecule of a molecular weight of 300 would take about an hour, and a protein would take one or more days to cover the distance.

Morphogens vary not only in their speed of diffusion, but in their ability to penetrate cells with their lipid membrane barrier. The small volatile signal substances go through the membrane without difficulty, as do the steroid molecules that are soluble in the lipid layer. But both proteins and smaller molecules that are not lipid-soluble cannot penetrate a cell and must rely on joining specific receptors on the cell surface to get their message through.

There is in this molecular size span a corresponding range of response mechanisms. In the case of the gases, their way of stimulating may be very general. This is particularly evident for ammonia and carbon dioxide where, once they enter the cell they increase or decrease the pH of the cell, respectively, and no doubt in many instances the appropriate response is achieved by this change in acidity. In the case of oxygen, for aerobic organisms there is a natural avidity for it; it is required for the internal engine, and an increase in oxygen will mean an increase in metabolic activity. There remains the possibility that in some instances there are specific receptors for these particular small molecules, but there is no evidence of such at the moment. In plant development, ethylene, another volatile signaling substance, is known to stimulate various aspects of growth and fruit ripening, and in this case there are specific receptors.

It is also true that middle-size morphogens will attach to specific receptors, either at the surface of or inside the cell. The larger protein morphogens, because of their increased complexity, will be paired with an even greater diversity of corresponding receptor proteins.

What we see as we go from small to large morphogens is a spread from more general to increasingly specific responses. This change is accompanied with a decrease in ability to act at a distance. It is less certain that the larger the morphogen, the more likely it will be associated with a large, complex organism. However, it is true, as we will see in the next chapter, that in many of the primitive multicellular organisms small morphogens do play an important role in development.

There is another important principle. Both a selection for the signals and for the receptors must have occurred.

No doubt they often arose in conjunction, by coevolution, but sometimes one or the other may have been selected for independently. For instance, it is common to find that one morphogen will have numerous roles. The plant growth hormone auxin is known to have multiple effects on development: it can cause cell elongation, cell division, stimulate root formation, and inhibit root growth, and it is one of the stimuli for leaf fall in deciduous plants. Another example, also involving a small molecule, is the different roles of cyclic AMP in cellular slime mold development. As in the cells of animals, cyclic AMP plays a major role in signal transduction and carries the message from activated surface receptors to middlemen molecules within the cytoplasm called kinases, which in turn stimulate cell differentiation. In addition, in some species it plays a vital extracellular role as well: it is the molecule that is responsible for attracting the cells to central collection points, for the individual amoebae orient in external gradients of cAMP. It is tempting to assume that during the course of the evolution of development, substances that are present in or among cells serving one function have been co-opted to serve as the signal molecule for a new function (see Gerhart and Kirschner, 1997, for other examples).

This kind of molecular opportunism is also open to the receptor proteins. For instance, they can evolve different molecular forms that will respond differently to the same signal molecule. For one receptor, the signal is translated into turning something on, but a sibling receptor will turn something off when stimulated by the same signal. For instance, in higher plants, auxin will stimulate the initiation of root growth, but also inhibit root growth if the cuttings are kept in the auxin solution. It is also true for these and many other examples that the same receptors will be responsive to the amount of signaling substance: over a threshold amount, the reaction will be the reverse of that below the threshold. So we have a further variable in the concentration of the signal molecules. There is also the important matter of the distribution of the receptors, which greatly affects the way development proceeds, and the matter of how their distribution is controlled in the first place—matters that are central to the whole phenomenon of pattern formation.

## Social Insects

The main thesis of this essay is the idea that, by looking at the onset of extracellular signaling in the first multicellular organisms, one might gain insight into the first principles of development in multicellular organisms. There is a remarkable parallel to be found in the extra-organism signaling in social insects, for there, also, are signals between individuals that serve to give a kind of unity to the whole colony. Perhaps we can learn from what is known from the way they manage integration and apply that knowledge to primitive multicellular organisms.

The biggest difference between insects and cells is that cells adhere to one another; they are in some form of intimate contact and constriction. Social insects, on the other hand, roam free and only periodically touch one another with their antennae. There are, however, a number of most interesting similarities. For instance, both have a division of labor, the proportions of which are controlled by signal systems, and both show striking instances of what I have just discussed, where the same signal will produce different responses in different parts. Let me give examples of both.

The idea that with a division of labor the proportions of the differentiated parts are consistent and controlled from generation to generation is obvious, and this is found in the simplest multicellular forms. In the next chapter we will see examples from cyanobacteria, the green alga Volvox, and the cellular slime molds. Let me now give a good example from the social insects. The proportions of different ant or termite castes are regulated by chemical signals, or pheromones that control the development of juveniles, so that the ratios of the different castes remain constant. These chemical messages, distributed from one individual to another by touching, are often inhibitors that repress the development of a particular caste. For instance, Light (1942–1943) showed that if the soldiers were removed from a termite colony, juveniles would, at the next molt, metamorphose into soldiers until their original ratio in the colony was restored. However, if those juveniles were fed a paste made of dead soldiers, no such metamorphosis occurred; the inhibitor in the paste prevented the switch in the developmental pathway. So, even though the individual workers are rushing about independently, they keep touching one another and in this way distribute the inhibitor pheromone.

There is also an example of different responses to the same signal, as was shown for a species of ant by Moser and his co-workers (1968). Low concentrations of an alarm substance will cause the workers to be attracted to its source, while higher concentrations produce the opposite response of fleeing. Concentration threshold effects of this sort are also known for developmental morphogens.

One process that initially seemed very different between multicellular organisms in general and social insect colonies was that the latter seemed to have fewer signals. At first I thought this was because multicellularity had a more ancient history than insect societies and that with time the signal response system became more and more complicated by selecting for a network of modifying and double-assurance signal-response agents. But there are two things wrong with this argument: the first is that social insects, while more recent than the origin of multicellularity, are nevertheless very ancient; the other is that as we delve more into social insect signaling systems, we find that they too are more complex than initially imagined. For instance, it used to be thought that in honeybees the queen controlled worker behavior by producing one or two inhibitor pheromones; now it is known that in fact she produces a complex of chemicals (Winston and Slessor, 1992). Again, we will see parallels among simple multicellular organisms that have existed for many millions of years, time enough to make their signal-response systems more complex, a matter to be discussed in more detail presently.

Like the matter of size, cell signaling is also one of the foundation pillars of both evolutionary and developmental biology. They are of pervasive significance every step of the way. Now we are ready to combine them with the different experiments in multicellularity surveyed in chapter 3 and see what insights they will give us into the ground rules governing development.

# 6 *The Basic Elements of Multicellular Development*

WHAT WE ARE seeking are the first principles of development. As I have pointed out, there are two ways of achieving this: one is by mathematical modeling, and the other is by looking at the beginning of multicellular development. Both approaches involve a considerable burden of hypothesis, for we were not there in the beginning. The simple organisms we see today have an ancient history, and an enormous time span has passed since they first arose—an almost infinite number of generations, time enough for many adjustments and alterations to cloud the way it was in the beginning. Let me nevertheless try to piece together the puzzle.

Previously I have laid stress on the variety of ways multicellularity itself may have arisen in early evolution; now I want to ask what are the common elements of development in those early organisms so that perhaps we can view development in its utmost simplicity. The underlying purpose of this essay is to find satisfying simple explanations for the mechanism of development, based on the assumption that those beginnings must have consisted of only the bare essentials and therefore permit us a peek at the fundamental properties of development—development without frills: ur-development.

My plan is to examine what are the underlying properties of all developments by looking for what might have been the first signs of coordinated development that arise with the advent of multicellularity. This will be followed by specific examples examined in some detail.

Perhaps the most fundamental property of any development—including the development of unicellular organisms—is polarity. All organisms have some kind or internal orientation, some inner directional quality. Here I will consider two kinds. In one there appears mainly an alignment of the cells—they alone have the polarity, and there is no "head," no central, dominant region within the group of cells; for convenience I will call it "headless polarity" (although it would be equally appropriate to call it "tailless" polarity). The second kind of polarity is the more common and familiar one in which there is a clear head-tail orientation, and for this I will use the conventional term "anteroposterior polarity."

Another fundamental element of development whose origin is of great interest is pattern. With it come two attributes: (1) cell differentiation with some sort of simple division of labor, and (2) a way to produce consistent proportions, or patterns, of the differentiated cell types from one generation to the next. We naturally think of these properties as being the essential parts of the development of higher plants and animals; here we are interested in how they arose in early evolution—what is their simplest manifestation.

One way we can look for those beginnings is to examine the primitive multicellular organisms that exist today, fully realizing that those living at this time have been in existence for many millions of years; they are only the descendants of the real pioneers. And during that incredibly long time span they might well have found complex ways to achieve simple developments. At the moment, there is no solution to this problem, although it is conceivable that ultimately we might be helped by studying the diverging cell lineages using the methods of molecular phylogeny.

## The Origins of Polarity

### *Headless Polarity*

Perhaps the easiest example of this comes from simple filamentous growth as is common among eubacteria, cyanobacteria, and algae. Among the green algae, *Spirogyra* is an obvious candidate; it consists of a row of cells with no head or tail. There are innumerable similar examples not only among the algae, but in the cyanobacteria as well. One cannot know if they are primitive and represent a primordial state, or whether they are descendants of forms with antero-posterior polarity; but perhaps that is less important than the fact that there clearly are niches for this kind of filamentous structure, for they are so common today.

Another class of examples comes from organisms where the cells are not in contact but are moving in the same direction. There are some bacteria that consist of motile rods that move in parallel formation to form swarms. Frequently, these swarms will move in leaderless circles. These patterns show clear symmetry, clear polarity, although it is not evident to what extent they can be considered stages of development, for in general they are feeding cells that swarm in this fashion. By being in groups they can feed more effectively, for large numbers of cells can secrete a high concentration of digestive enzymes into their immediate environment and in this way make bigger prey palatable. In the myxobacteria these swarms eventually do come to a head, a center, to form a fruiting body, so they go from simple alignment to an antero-posterior configuration all in one life cycle.

Headless swarms are by no means confined to bacteria. They are also found in the curious eukaryotic marine parasite of eel grass, *Labyrinthula,* which forms a feeding net of amoeba-like cells that crawl up and down within a tubular meshwork of slime that they themselves have secreted (fig. 18).

Let me add parenthetically that such headless orientations are also found among groups of higher animals, such as schooling in fish and trail following in ants; it is not a phenomenon found only in the dawn of early development. Each of the organisms forming the swarm are polar—they have heads and tails—yet there is no leader; all of them are followers. There is considerable current interest in how fish schooling and other oriented movement among motile animals can be achieved with such remarkable precision and coordination (for example, Deneubourg and Goss, 1989).

In all these lowly, multicellular cases there is some communication between the swarming cells. This has been ex-

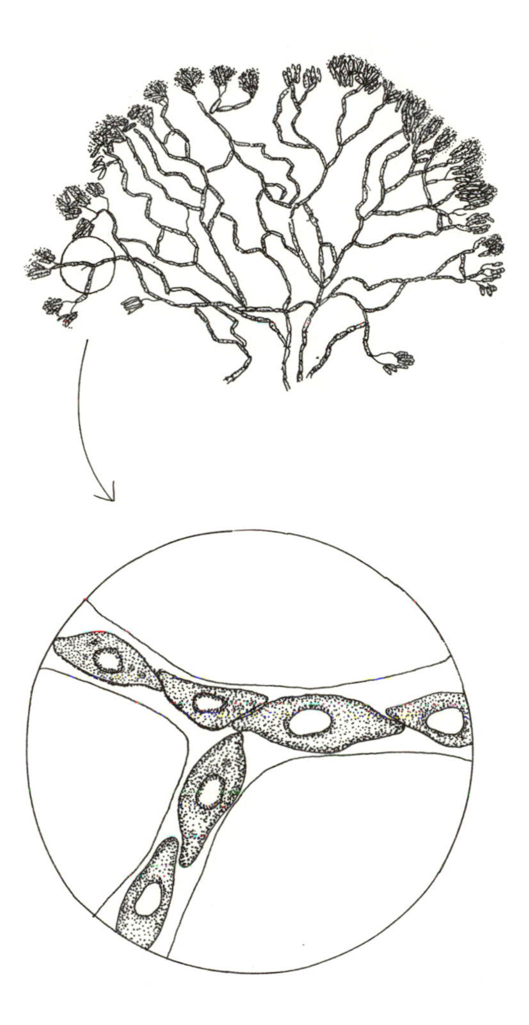

Fig. 18. (*Top*) The feeding edge of a colony of *Labyrinthula*; (*bottom*) a high-power view of a portion of the net showing the individual cells. (From Bonner, 1974.)

amined in some detail by Kaiser and his co-workers in myxobacteria, where they find that starvation first causes the cells to produce a factor that signals high cell density, and this is followed by another factor that stimulates the cells to move together as a swarm that ultimately migrates to a suitable location for the formation of spores (for a general review of experiments on myxobacteria, see Dworkin and Kaiser, 1993). In the case of cellular slime mold aggregating amoebae, they sometimes come together in a circle or a torus (like army ants following one another in a circle—see Schneirla and Piel, 1948). All the amoebae are moving in the same direction and have adhesion molecules that allow them to cling to one another; only at some later moment will they become unstable, break apart at some point in the ring, and have a clear front end (see Shaffer, 1957).

*Antero-Posterior Polarity*

There are many factors, both internal and external, that will lead to antero-posterior polarity. One of the key influences is the immediate environment; it may play a decisive role. I will give a variety of examples.

To continue with simple filaments, the green alga *Ulothrix* is similar to *Spirogyra* except it has a basal cell which attaches to the substratum—it is a holdfast cell (fig. 19). Here the axial polarity does not start at the "head" but rather at the foot of the filament, and obviously it serves the purpose of keeping the filament fixed in one place. So we have two ecological strategies here: one where the alga is best off catching the sun by moving along with the current, and the other where it is best off fixed in one place.

Fig. 19. The growth of filaments of the green alga *Ulothrix* showing the holdfasts. (From W. H. Brown, *The Plant Kingdom*, Ginn, 1935.)

It is impossible to know which came first; all we do know is that there are two different ecological niches, and there are many algae that fill each of the two. It is true that if one looks at prokaryotes, the cyanobacteria have not invented holdfasts, but they have found other ways of attaching to one place, such as producing a sticky jelly which fixes groups of filaments in one spot. In other words, they too have filled both niches, although in different ways.

In photosynthetic filaments such as we have discussed, any cell in the filament can divide, for the cells feed on the sunlight. There are exceptions, however, where there is apical growth, and this is invariably true for the non-photosynthetic fungi. In these instances, the antero-polarity does not just start in the toes, so to speak, as it does in

*Ulothrix*, but there is a head as well, the growing tip. The reasons for this difference are no doubt mechanical, that is, the mature cells with stiff walls cannot divide and require that one region of younger, softer cells be the growth zone, or meristem. There might be adaptive reasons as well, although one can hardly be sure.

In nonfilamentous colonies the environment can also play a significant part in producing an antero-posterior polarity. In its simplest form, as previously mentioned, it has been suggested that the archeobacterium *Methanosarcina* lives in an anaerobic environment, and its organization consists of little more than a heap of cells, yet it has the advantage of being protected against an influx of lethal oxygen. Should such an influx occur, at least the central cells will be safe, buried and insulated in the interior. This is more an illustration to show how the environment might have produced conditions that would favor selection for antero-posterior polarity, for the end result—a mound of archae cells—hardly qualifies as a well-organized multicellular organism.

## Polarity in *Fucus*. Environmental and Cell-Produced Stimuli

A classic example of the development of antero-posterior polarity is the development of the egg of the brown alga *Fucus*. Admittedly, it is a large and highly developed alga, but it has been advantageous material for experimental work, and from it we can learn many of the basic principles

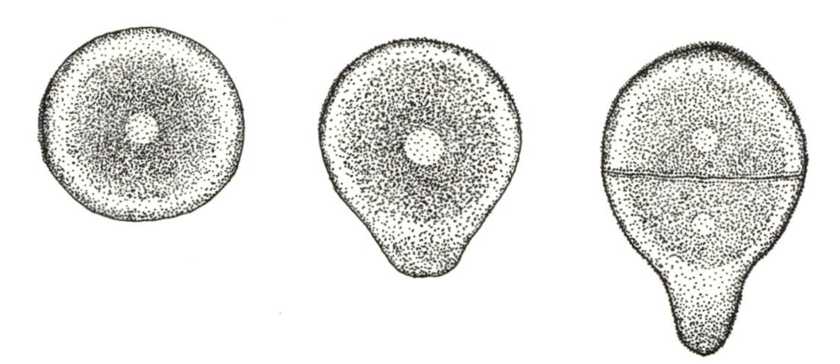

Fig. 20. Rhizoid formation in the egg of the brown alga *Fucus.* The egg is initially symmetrical around a point, with the nucleus in dead center. First, the rhizoid appears as a bump, and then the first cleavage plane forms at right angles to the axis of elongation. (From Bonner, 1974.)

of how the environment can impart antero-posterior polarity to an organism.

*Fucus* eggs have the unique property of starting as perfect spheres with the nucleus in the dead center; initially, it has no antero-posterior polarity at all. Polarity appears with the onset of development: first, the beginning of a rhizoid appears as a protrusion at one end, followed by a cleavage plane at right angles to the newly acquired long axis (fig. 20). Once established, this antero-posterior polarity remains fixed throughout the entire development of the plant.

How the environment determines the direction of this polarity in *Fucus* and related algae is a story with a long history and has been discussed in detail in many places (for example, see Jaffe, 1968; Quatrano, 1978). Presumably it is important that the rhizoid end points towards the ocean

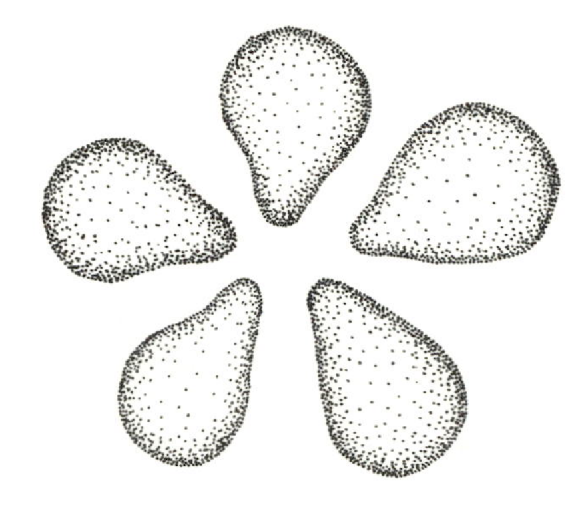

Fig. 21. Developing *Fucus* eggs showing the "group effect" in which the rhizoids point towards the center of the group. (From Bonner, 1952.)

floor for it to develop into a holdfast, and the front develops upward towards the light to allow photosynthesis. The nonpolar eggs are sensitive to light, and the rhizoid will form on the dark side. The eggs are also sensitive to chemical influences, as was first observed in the so-called group effect, where, if groups of eggs are near one another, their rhizoids will form inward, towards one another (fig. 21). This phenomenon is the result of a chemical they secrete called rhizin (probably the plant growth hormone indole acetic acid) which stimulates growth. Its concentration is highest in the center of a group of cells and therefore stimulates rhizoid growth at the centripetal end of the eggs. Also, rhizin would be most concentrated on the side of the egg touching the substratum, again initiating the correct orientation of the future plant. The eggs also secrete an inhibitor of rhizin (antirhizin), which plays a role in maintaining the proper distribution of rhizin's growth-promoting gradient.

Much is known about the mechanisms of how light and rhizin achieve their effects, but for this discussion the important point is that the environment can, in two distinct ways, control the polarity of the developing egg. Furthermore, both ways lead to the same goal: to see that the orientation of the early embryo will be optimal for the survival of the adult.

### Polarity in Hydroids. Environmental and Cell-Produced Stimuli

In many organisms, the environment can determine the antero-posterior polarity in another way: by an oxygen gradient. This notion goes back to the early work of C. M. Child (reviewed in 1941), who championed the idea of metabolic gradients. He noted that in all developing organisms, the antero-posterior polarity is invariably tied to a metabolic gradient, with oxygen consumption highest at the anterior end. Clearly, one way that this can be initiated is by having an embryo in an environment in which there is more oxygen surrounding one end than the other. The high oxygen can stimulate the metabolic activity at that end, in this way setting up an oxygen-consuming gradient and establishing the antero-posterior polarity. According to Child's evidence, once an oxygen-consuming gradient is in place, it is self-maintaining. The high metabolism in the anterior region is autocatylic and dominates by taking the lion's share of the substrates to feed its metabolism.

An example of how such a system operates is given by the old experiments on the regeneration of isolated pieces of the stem of the hydroid *Tubularia.* Admittedly, hydroids are hardly primitive multicellular forms, but as in *Fucus,*

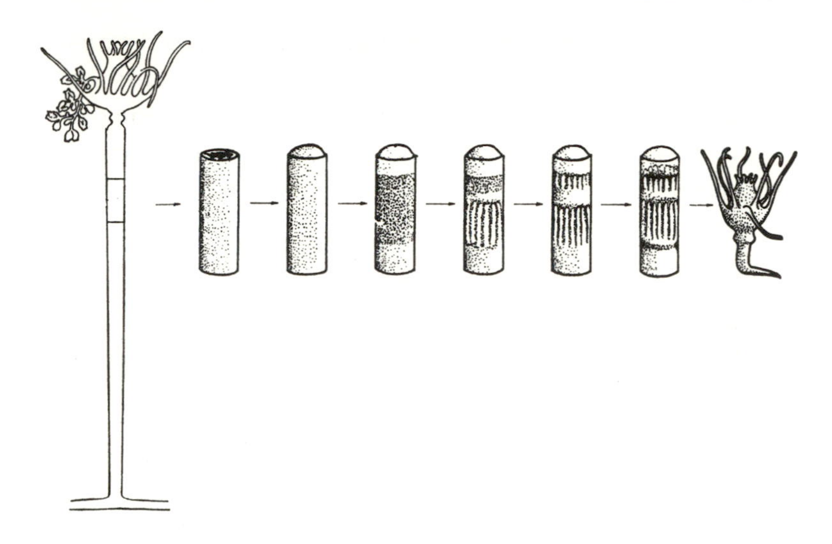

Fig. 22. An adult polyp of *Tubularia* and the steps in the regeneration of a cut portion of its stem. (From P. Tardent, *Biol. Rev.* 38[1963]: 293–333.)

they perfectly illustrate how oxygen can play a role in establishing polarity.

The initial experiments were done in the early part of this century by T. H. Morgan (1903). He stuck cut pieces of *Tubularia* stem into the sand, and no matter what the original polarity of the stem segment, invariably the end in the free water became the new anterior end with a mouth and tentacles (the hydranth end) (fig. 22). Morgan postulated that the free end had access to more oxygen than the end buried in the sand, and this difference was enough to reverse the polarity of the piece of stem. Much later this was elegantly confirmed by Miller (1937), who devised a small, clever, two-chambered aquarium in which the pieces of stem were sealed into holes in the partition, with their ends protruding into the two chambers. By varying the oxygen concentration in the seawater in the two chambers, Miller was able to confirm that the hydranths formed on the side with the higher oxygen tension regardless of their original polarity.

It turned out that oxygen was only half the story; the hydroid produced an inhibitor, and its presence prevented a hydranth from forming. So, in Morgan's sand experiment there was not only less oxygen around the ends stuck in the sand, but more inhibitor, and both these factors prevented the buried end from becoming a head end. As in *Fucus*, there is an internally generated chemical mechanism that reinforces the external information.

I should add that all these polarity-inducing phenomena that I have described for *Fucus* and *Tubularia* provide textbook cases for applying the principles of reaction-diffusion models (see, for example, Gierer and Meinhardt, 1972; Meinhardt, 1982; Murray, 1989). In fact, Alan Turing, in the 1952 paper that began the whole idea of reaction-diffusion models, used hydroids as an example. I will return later to the matter of mathematical models in developmental biology.

## The Beginnings of Cell Differentiation

After the establishment of polarity, cell differentiation follows. One of the ways differentiation could arise is by having differences in local environments within a collection of cells, and those compartmentalized environments would lead to differences in the cells. I have argued elsewhere (1996) that if such a phenotypic event occurred repeatedly for many generations, any gene mutation that made the same cell change would become fixed in the genome. I called this "gene accumulation" and argued that such genetic fixation would arise because there was no selection pressure against genes that mimicked a phenotypic event that favored the fitness of a developing organism. In the case of early differentiation, the asymmetrical environment led to a difference in the surroundings of some of the cells due to their location, and those differences became progressively fixed by gene accumulation.

I fully appreciate that my "gene accumulation" is a gross oversimplification, albeit a useful shortcut. As West-Eberhard (2000) has shown in elegant detail, any shift in the genetic constitution of an organism involves a complex set of factors, which involve the degree of phenotypic plasticity and the modification of expression of various genes, especially those involved in regulation. To encompass these sets of changes she uses the more general concept of "genetic accommodation" to include all the interactive changes in the genome that could stem from persistent, environmentally induced effects. The process is progressive, and there may well be many genes and gene interactions involved in the production of a cell-generated stimulus that can act in the same way as the stimulus provided by the environment. Such an internally produced signal could be a morphogen, and along with it there must be a response mechanism to the signal, all of which eventually becomes incorporated into the genome. (See West-Eberhard, 2000, for a detailed discussion of genetic assimilation and the Baldwin effect and their relation to genetic accommodation.)

In both *Fucus* and *Tubularia*, for example, one might imagine that first the environmental cues produced the polarity and that much later the production of internal inhibitors, or morphogens, reinforced the same polarity induction.

## Differentiation and Pattern in the Cyanobacteria

Another example might be argued from the formation and the pattern of the distribution of heterocysts in the filaments of cyanobacteria. Furthermore, in this case we are dealing directly with a primordial multicellular organism. Becoming multicellular opens the gateway for all sorts of remarkable innovations that would be impossible for single cells. By being larger and made up of numerous cells, organisms can not only have a division of labor but can respond to their environment in new and sensitive ways, all adaptations that have led to their success.

Cyanobacteria, which are known from the fossil record to be a very ancient group, have two biochemical functions that are imiscible, leading to a division of labor. They are photosynthetic organisms and therefore produce oxygen in the presence of sunlight, and at the same time they must fix the free nitrogen from their immediate environment, an absolute necessity to make their proteins and all the other component molecules in their body that contain nitrogen. Since nitrogen fixation can only take place in the

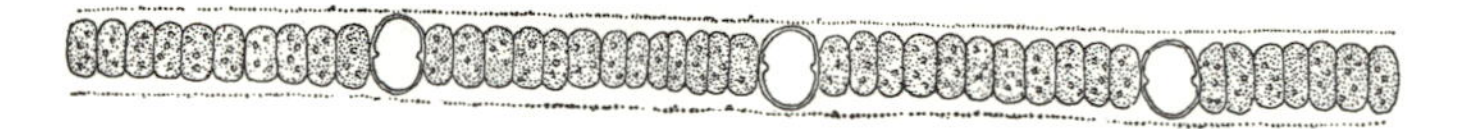

Fig. 23. A filament of the cyanobacterium *Nodularia,* a relative of *Anabaena,* showing the even spacing of the heterocysts. (From G. M. Smith, *The Freshwater Algae of the United States,* McGraw-Hill, 1933.)

total absence of oxygen, the two processes essential for their existence cannot occur simultaneously in the same place. Many species of cyanobacteria have solved the problem by doing their photosynthesis in the daytime and their nitrogen fixation at night, but here I am more interested in a second and more sophisticated solution, where some of the cells in the filament become specialized for nitrogen fixation.

These so-called heterocysts are clear cells, lacking chlorophyll, and have thick walls that prevent the oxygen from neighboring cells from getting into their inner machinery. By this cellular division of labor, cyanobacteria can take in energy from the sun and use it to make nitrogen compounds while simultaneously pulling in nitrogen from their surroundings. The nitrogen products are passed along to the photosynthesizing cells through the pores at the end walls of the heterocysts; all the cells therefore benefit from their work, and they in turn receive the nutrients they need from the neighboring green cells.

Over and above these remarkable facts, it is known that in some species the heterocysts are perfectly spaced along a filament of cells (fig. 23). Wilcox et al. (1973) have shown that the heterocysts in *Anaboena* give off an inhibitor that diffuses along the filament, which prevents any cell for the limit of its effective diffusion from turning into a heterocyst. So, despite the fact that cyanobacteria are early-evolving prokaryotes, they have both a cell differentiation into two cell types (in fact three, because some cells become resistant spores for surviving hard times), and this differentiation is organized into a regularly spaced multicellular pattern. I should add that the molecular genetics of this system have been examined by a number of workers and, considering how long the cyanobacteria have been in existence, it is not surprising that the genetic control mechanisms are remarkably complex (for reviews, see Haselkorn, 1998; Yoon and Golden, 1998). None of these achievements could have arisen in a unicellular cyanobacterium.

How did this bit of primitive organization arise in the beginning? One can only speculate. Let us assume that originally the terminal cell of a filament is partially buried in the bottom of the pond or ocean, and therefore is surrounded by less oxygen and less light than the other cells. Indeed, some species have their heterocysts only at one end of their filaments. Should such terminal cells be in a zone of reduced oxygen and light for many generations, genes might accumulate that favor those cells becoming nitrogen-fixation factories. The innovations would be the disappearance of chlorophyll and the formation of an impenetrable cell wall, with connecting pores to neighboring cells to pass on the nitrogen compounds they have manufactured. Once the genes necessary to produce heterocysts are in place, they might produce an inhibitor, a morphogen, that prevents heterocyst neighbors from following suit, thereby creating a consistent pattern in which the heterocysts are strategically placed on the filament. This could be a common way cell differentiation originates.

## Differentiation by Cell Lineage

Another intriguing possibility has been suggested by L. W. Buss (1987). He points out that a new cell line might arise by mutation in an early multicellular organism, much like cancer cells, and that it will be in competition with the ancestral cell line. If the mutant cells behave as a cancer and they take over, this will obviously not be a strategy that will lead to long-term reproductive or evolutionary success, but rather to catastrophic extinction. On the other hand, the initial mutation and the ones that follow it could reach an equilibrium where neither the new cell line nor the parental one disappears; they remain in some sort of balanced harmony.

This is exactly what is known to happen in the cellular slime mold *Dictyostelium mucoroides*, which is a common inhabitant of soil. M. Filosa found that with mass spore transfer, our cultures contained a mutant that hardly makes spore cells at all when it is grown alone, yet when it is mixed with the wild type, the phenotype is exactly similar to the wild type (Filosa, 1962; Buss, 1982, 1999). In other words, the mutant is a cheater, or in Buss's way of putting it, an auto-parasite, and it manages to remain in a fixed ratio to the wild type cells so that it can perpetuate its career of cheating.

Buss points out that such cell lineage competition, with its controlled cell type ratios, could be a beginning of cell differentiation. If the organism is asexual it can remain in this condition perpetually, or at least until another mutant cell line appears. On the other hand, if sexual reproduction is to occur, then the genes of both cell lines must in some way become incorporated into one cell, into the gametes, perhaps by the lateral transfer of genes. Once this is established, the controlled differentiation of the two cell types can be passed on to the next generation through sexual reproduction.

In the next chapter I will go into much greater detail—using cellular slime molds—to illustrate a primitive case of polarity and pattern formation. In this particular example we have a considerable amount of experimental detail on how polarity and pattern arise.

# 7 *Development in the Cellular Slime Molds*

My plan here is to illustrate my main points by examining how they apply to one organism. I want to use the cellular slime molds to show how one can look at their development from the three points of view: the biological, the molecular, and the mathematical. By going into one case in some depth, we can see how the three approaches dovetail. Together they help us to understand not only how multicellular development arose in the first place, but how all three ways of looking at the matter help us to understand development itself.

## The Biological Approach

I have already described the cellular slime molds as organisms that became multicellular by aggregation and are among those that achieve multicellularity on land. Growth involves separate, single amoebae feeding on bacteria in the soil, and once they have finished the local supply of food, they stream together to form a multicellular organism (fig. 9). In this way they totally separate, in time, their growth phase from their multicellular morphogenetic phase.

In *Dictyostelium discoideum*, the species most studied, the aggregate turns into a migrating slug containing many thousands of cells that not only move in a directed fashion, but are amazingly sensitive to their environment, apparently so that they can produce their spores in a spot for optimal dispersal. The smaller group of anterior cells of the slug are clearly different from the posterior cells; the former become the stalk cells, and the latter the spores. They fruit by having the slug stop its forward motion, right itself, and the anterior amoebae begin to become vacuolate at the very tip, forming a cellulose cylinder onto which they keep cramming themselves in a reverse fountain movement, ultimately causing the whole cell mass to rise into the air as the tip keeps elongating. The posterior amoebae are pulled up as though they were in a bag, and each one becomes encapsulated to form a spore.

The whole cycle takes about two to four days in the laboratory, and the slugs as well as the fruiting bodies are on the order of magnitude of a millimeter. Their size is utterly dependent on how many amoebae entered an aggregate, and, as discussed earlier, the relative number of stalk cells to spores is proportional over a considerable size range. In other words, their pattern is regulated. Let us now examine the matter of how the polarity and the pattern are established.

## Oxygen and Polarity in the Cellular Slime Molds

That oxygen gradients can play a significant role in establishing polarity in these social amoebae was first demonstrated by Sternfeld and David (1981b). They submerged spherical balls of amoebae in agar—amoebae that had been stained with the vital dye, neutral red. The stain has the property of darkly coloring the anterior prestalk cells (cells destined to become stalk cells), while the posterior prespore cells can be clearly distinguished because they stain lightly. Initially the balls of mixed cells embedded in the agar are uniformly stained, but soon those near the surface, where they have greater access to oxygen, turn dark red, while those in the interior are pale. No doubt two well-known phenomena can account for this color distribution: cells in the presence of high concentrations of oxygen become converted to prestalk cells; but also those cells that have prestalk characteristics but lie in the interior of the ball will migrate up the oxygen gradient to join the peripheral band of dark cells.

### *Capillary Experiments*

An effective way of showing the role of oxygen in establishing polarity is to draw aggregated amoebae up into a very fine capillary with air on one side of the cells and mineral oil on the other (Bonner et al., 1995). In a minute or two, the cells at the air end will form a sharply defined zone of amoebae with prestalk characteristics. These are characterized by the anterior cells being highly motile compared to the posterior cells, and they are able to synthesize a protein that is characteristic of prestalk amoebae. In further studies it was possible to establish that the length of this anterior zone is dependent on the concentration of oxygen in the bubble near the cells; the sharp division line appears to be caused by a specific threshold of oxygen concentra-

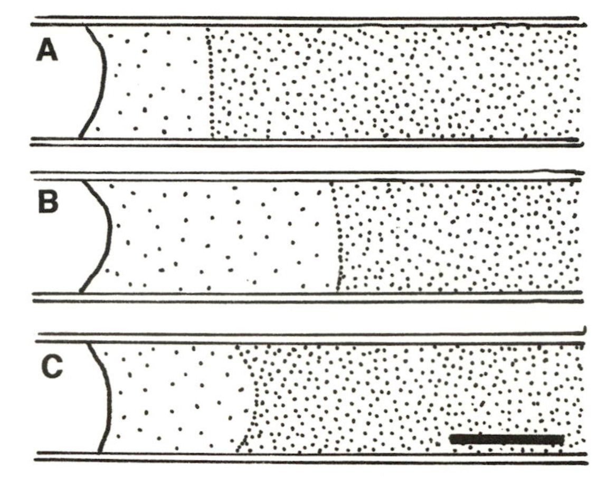


Fig. 24. Increasing the oxygen concentration increases the length of the anterior zone in slime mold cells drawn into a glass capillary tube. (A) Air in the space on the left. (B) Twenty minutes after 100% oxygen has been added on the left. (C) Seven minutes after putting air (21% oxygen) back in the space on the left. Bar = 100μm. (Drawn from a photograph in Bonner et al., 1998.)

tion, and any change in that concentration is rapidly accompanied by a shift in the position of the division line (Bonner et al., 1998; Sawada et al., 1998) (fig. 24). Clearly, the oxygen has not only established the direction of the antero-posterior polarity, but has done so with extraordinary rapidity.

At the same time we found another phenomenon of considerable interest. If an air bubble was placed at both ends of the cells in a capillary, within a minute or two each end showed a similar zone of active anterior amoebae. If the capillary was of absolutely even bore, this pattern remained stable for a number of hours, until the cells slowly died (fig. 25). On the other hand, if the capillaries were hand drawn, and as a result had an insignificant, imperceptible taper, the band in the ever-so-slightly larger end persisted, while the anterior zone at the smaller end gradually became smaller and smaller so that it was barely visible after a couple of hours. This experiment raises all sorts of interesting questions to which we do not know the answers yet. However, it does tell us that the cell mass is extraordinarily sensitive to minute differences in oxygen concentration and can invariably choose the larger end as the winning end, even though we are unable by eye (under the microscope) to detect any difference in size between the two ends of the capillary. What we do not understand is how one end can dominate the other: what sort of signals can pass along the column of cells so that one end can dictate supremacy to the other. What we do know is that oxygen is probably the trigger and that the establishment of the polarity of a cell mass is ultrasensitive to an amazingly shallow oxygen gradient.

### *Oxygen and Normal Development*

If we think of these facts in terms of normal development, the aggregation of the amoebae results in a mound of cells, and invariably the tip, which becomes the anterior end of the migrating slug, forms at its apex. Clearly, the cells at the top will be exposed to more oxygen from the air than the cells at the bottom of the mound, in this way establishing the initial polarity. One might also be able to argue that the apex of the mound would be exposed to slightly

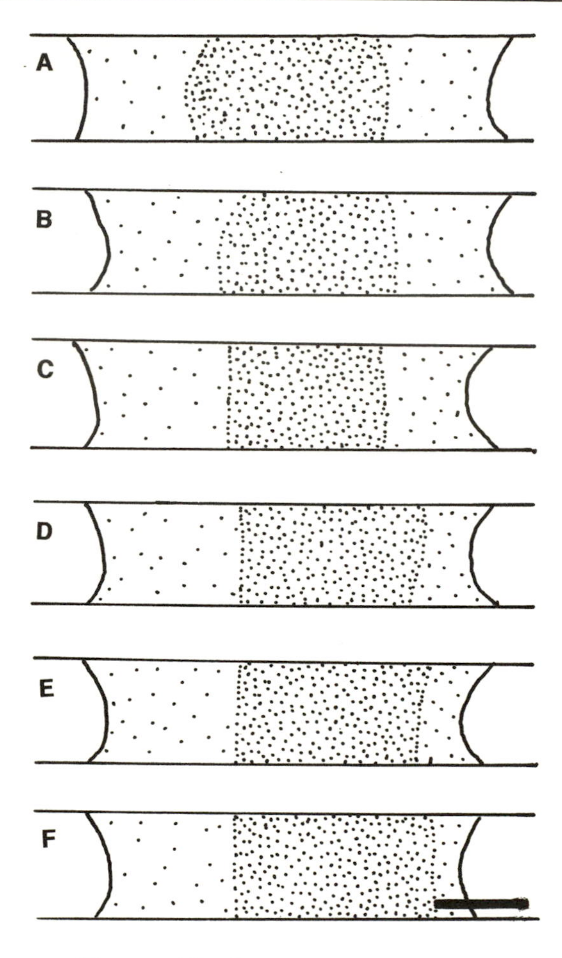


Fig. 25. Slime mold cells in a capillary with air at both ends. The capillary has an imperceptible taper, and the larger end eventually suppresses the zone of active cells at the smaller end. (A) Anterior zones form at both ends after about 1–2 minutes. (B–F) Images gather at 60-minute intervals thereafter. Bar = 100μm. (Drawn from a photograph in Bonner et al., 1998.)

more oxygen than its sides, and this minute difference would be enough to place the tip at the very zenith of the mound. Furthermore, the slime sheath will be thinnest at the tip, which would also help oxygen to penetrate more readily there. So an ultrasensitivity to the oxygen gradient might play a significant role in normal development.

With these new facts in mind, and these thoughts on polarity, let us look at some old experiments. Of particular interest are those of Yamamoto (1977), who put slugs vitally stained with neutral red into small tunnels in agar in which they fit snugly. When such a slug reached the dead end of a tunnel, each cell within the slug starting at the tip did a U-turn so that soon the entire slug, with its anterior red tip, was going in the opposite direction. The dead end reversed the polarity, but instead of the posterior prespore cells becoming prestalk, the red prestalk cells kept their identity and individually doubled back. One might postulate that those anterior cells, when they hit the dead end, suddenly had more oxygen behind them than in front and reoriented in the opposite direction. There are two reasons one might suspect this possibility: one is that there would be a higher metabolic rate at the tip, which would deplete the oxygen faster at that sealed-off end of the tunnel; the other is that the slug will have oxygen in the gas phase at the rear, which is the open end of the tunnel, while the anterior end abuts the agar. I quite realize there are other possible interpretations; the main reason I find this one tempting is that the amoebae are so sensitive to small differences in oxygen tension, as we have just seen. Also, recall that Sternfeld and David (1981b) showed that prestalk amoebae in a clump embedded in agar oriented and moved towards high oxygen in a gradient. However, we also know that the slug tip is sensitive to ammonia and will orient in a gradient (Kosugi and Inouye 1989); perhaps both gases are acting together, with one attracting and the other repelling.

It is well known that cells can percolate through one another, either in opposite directions, as in this case, or in the same direction. There are many experiments of others that show the same thing, but years ago I plugged some anterior vitally stained cells into the posterior end of an unstained slug, and the stained cells wormed their way to the anterior end (1952b) (fig. 26). In today's terms, we imagine that they moved forward because they were particularly sensitive to the cAMP signals emanating from the tip.

Note that this is much the same story that we saw for hydroids, where high oxygen induced a hydranth when regeneration occurred in an oxygen gradient. Most notable were the experiments on *Tubularia*, where oxygen could reverse the original polarity of a piece of regenerating stem. Another parallel is to be found in some experiments with *Cordylophora*, where Beadle and Booth (1938) separated the cells into a heap, much like the mound formed by aggregating slime mold amoebae, and as they reorganized the hydranth formed at the top of the cell rubble. They argued that this was likely explained by the higher oxygen concentration on the upper surface.

### *Polarity Leads to Pattern*

The amoebae inside the capillaries, however, lack many of the properties of normal slugs; in particular, the posterior cells do not show prespore characteristics, and the cell

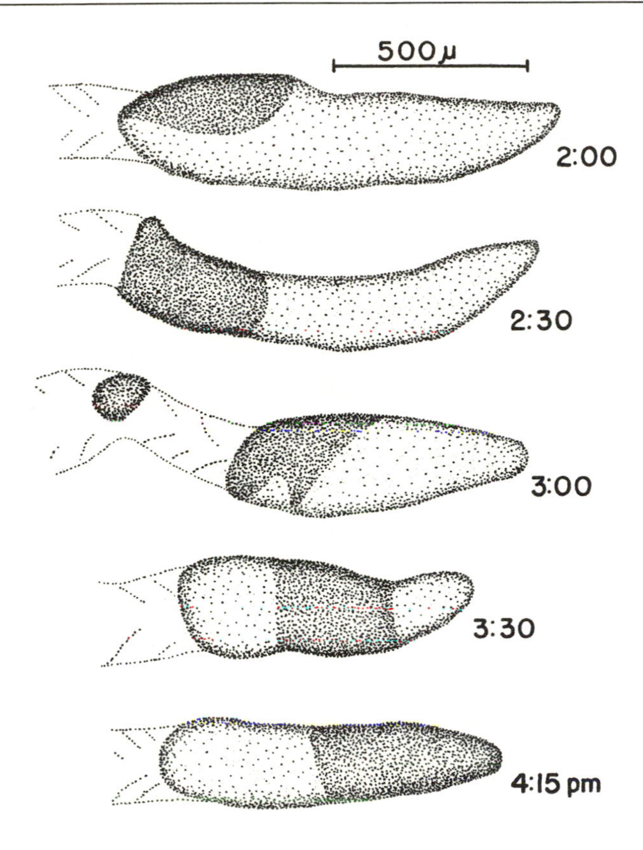


Fig. 26. Camera lucida drawings showing the rapid forward movement of a group of anterior cells from a vitally stained migrating slug that have been grafted into the posterior end of a colorless slug. Note that in the middle drawing a piece of the graft was lost. (From Bonner, 1952b.)

mass does not move a unit. By a stroke of good luck I found a way to produce occasional very small, flat slugs that are often only one cell layer thick. If capillaries containing cells are broken into pieces under mineral oil on a microscope slide, sometimes a small group of active cells will crawl out between the oil and the glass. These minute two-dimensional slugs are normal in many respects: they have antero-posterior polarity, they move forward, they have a slime sheath, they occasionally split in two, or twin, and they have a clear prestalk zone of very active cells and a prespore zone of conspicuously less active cells. One of the remarkable things about these flat slugs is that they are so small. They range from just over 100 amoebae to 2,000, while normal slugs range from 8,000 to 2,000,000 amoebae with a mean of approximately 500,000.

If these minislugs are observed in time-lapse video, where often all the amoebae are in a monolayer, one can see a big difference in the motion of the anterior prestalk cells and the posterior prespore cells. The latter move straight forward with military precision, like a flock of sheep moving down a narrow street. Each prespore cell appears to be moving forward by pseudopodial activity as the orderly mass advances as a whole (fig. 27).

The anterior prestalk cells, on the other hand, move far more rapidly and they go in all directions, as one might imagine gas molecules zipping about inside a balloon. We must remember that this anterior zone is responsible for the direction of the slug; it leads, and the posterior prespore cells follow (although they do so under their own power). If one examines the tip closely, especially when it turns, one first sees the most active region of cell move-

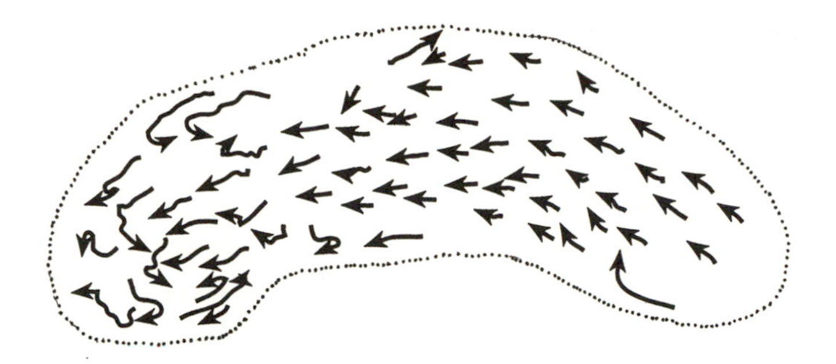

Fig. 27. The movement of individual cells in a two-dimensional slime mold slug showing that the anterior cells are more active and swirl about, while the posterior prespore cells move straight forward. The arrows follow individual cells from fifty frames of a video spanning six minutes in real time. The slug is 175 µm long and is made up of about 225 amoebae. (Drawn from a photograph in Bonner, 1998.)

ment, with its cells racing backwards and forwards, shifting to one side, and all the cells following it; there is no fixed set of cells at the tip.

I would like to argue that this is possibly an important, independently derived form of multicellularity, and from it we can extract some lessons. It is well known that the amoebae, not only during aggregation but also in the slug stage, are being attracted by gradients of cyclic AMP. Furthermore, it was established years ago that there is a gradient of extracellular cAMP secreted by the slug, highest at the anterior end and diminishing as one proceeds posteriorly (Bonner, 1949). No doubt this extracellular gradient is also present inside the slime sheath and the point of highest concentration is at the very tip. If amoebae move near that apical hot spot, they are stimulated to secrete more cAMP; it is a self-feeding, autocatalytic process. The key point is that this gradient is maintained in the surroundings of the cells and not within the cells themselves. Extracellular diffusion of morphogens has also been demonstrated in a number of animals; for instance, Gurdon and his co-workers (1998) show it in amphibian embryos.

It is also known that slime mold slugs not only produce ammonia but can be induced to turn by different concentrations of ammonia on different sides of the slug (Kosugi and Inouye, 1989). Furthermore, it is well established that ammonia inhibits the synthesis of cAMP. If one watches turning in the videos of two-dimensional slugs, one sees that the very active zone at the tip, where the cells are vigorously milling around and rushing in and out, moves to one side. It is not as I once thought that all the cells on one side move faster than the other, but it is the tip region that moves, and this seems to lead all the cells behind it in a new direction. This supports the straightforward hypothesis that when turning, the active, cAMP-high tip region is shifted to one side because there is more ammonia generated by the cells on the other side; ammonia governs direction by shifting the position of the tip. This ammonia can be produced by the cells themselves, or it can come from some external source.

### *Pattern Formation*

I think of pattern as something that follows directly from polarity. The latter implies solely orientation, while pattern is much more—not only the differentiation of parts, but also their arrangement in space in some kind of a consistent organization. It is conceivable to have a pattern without polarity, although this is stretching the point a bit. For instance, the arrangement—the spacing—of trees in a forest is strictly nonrandom; there are a number of mechanisms that cause this to happen, but there is nothing in a forest that could be called polarity. On the other hand, a forest is not an organism. The same arguments could be applied to the aggregation centers of cellular slime molds, where again there is a nonrandom spatial distribution of the aggregates, and therefore the fruiting bodies, yet there is no polarity.

One of the most important features of pattern in general, including pattern in the slime molds, is that consistent proportions are maintained from generation to generation. This is true for the limbs of a mammal, for the petals on a flower, and for the proportions of stalk and spore cells in slime molds. There have been numerous studies on the details of this proportionality, not only in slime mold fruiting bodies, but in the prestalk and prespore zones of the migrating slug (reviews: Bonner, 1967; Nanjundiah and Bhogle, 1995).

This consistent pattern is evident in the minute two-dimensional slugs described above in which the anterior prestalk cells show a totally different kind of cell movement from the posterior prespore cells (fig. 27). Furthermore, this is known not to be the lower limit for the control of proportions. In some experiments I did many years ago, I achieved a world's record with a fruiting body of *Polysphondylium pallidum* consisting of seven cells: three stalk cells and four spores (Bonner and Dodd, 1962) (fig. 28). So all the wonders of the control of development can oper-

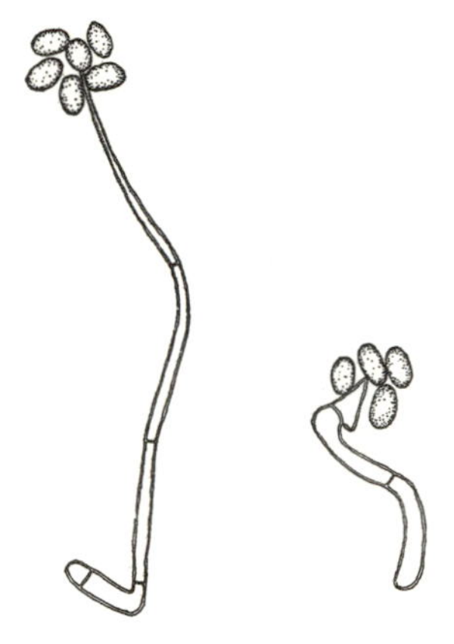

Fig. 28. Minute fruiting bodies of the cellular slime mold *Polysphondylium pallidum* consisting of eleven and seven cells, with a proportional distribution of stalk cells and spores. (From Bonner and Dodd, 1962.)

ate over a size range from seven to two million amoebae (although it has been pointed out to me by my colleague Ted Cox that any kind of diffusion explanation of proportions for seven cells presents formidable difficulties). If one thinks of all the morphogens that are even known or suspected to be involved in producing consistent proportions (to be discussed in the next section), it seems remarkable to me that the complex set of signals that must diffuse from both ends can operate over such a wide range of cell numbers, and even in two dimensions as well as the normal three. They began in an oxygen gradient as they started out from the broken capillary, but then as they glide along under the mineral oil (which is rich in oxygen) they are under a uniform oxygen tension, yet they produce and maintain two proportional zones (fig. 29). How they manage this is a matter to be discussed in the next section, where we consider molecular and biochemical matters.

The lesson we learn from cellular slime molds is that

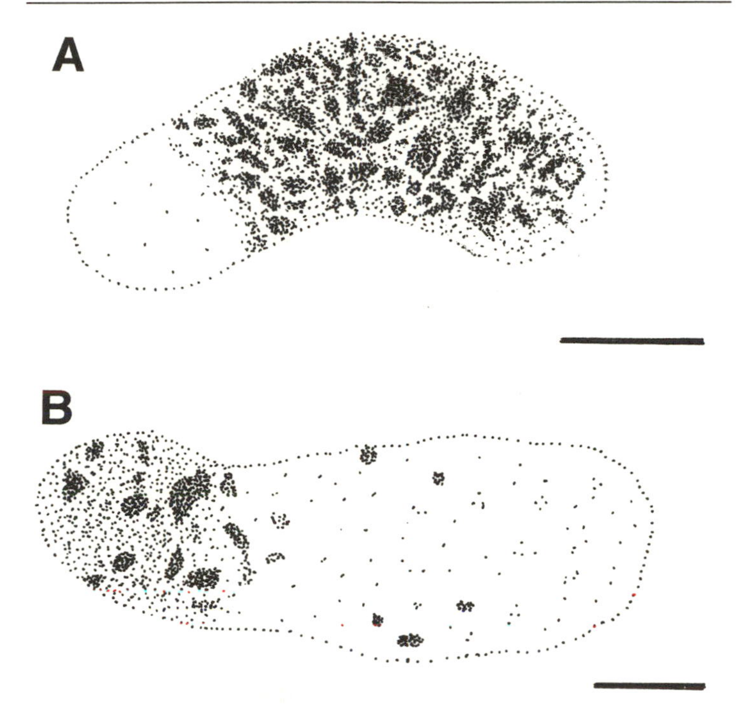


Fig. 29. Two-dimensional slime mold slugs in which a fluorescent label (GFP) has been attached to the promoter of (A) a prespore protein (*psA*; slug of about 500 amoebae), and (B) a prestalk protein (*ecmA*; slug made up of about 1000 amoebae). Note that the two zones are in the appropriate places. The scale bars are 50 µm long. (Drawn from a photograph in Bonner, Fey, and Cox, 1999.)

polarity arises first; it is something that can be set in place very rapidly, and it has an extremely delicate sensing mechanism. The pattern formation that follows uses the polarity as a guide, so that it is arranged in the right direction and therefore in an orderly fashion.

### *Social Amoebae Compared to Social Insects*

There are some interesting and instructive parallels and differences between social insects and slime molds. One involves polarity. We saw in slime molds that cAMP appears to be autocatalytic, which results in gradients within the cell mass. In ant colonies, the shape of an ant swarm again is to be found in the reactions between the ants and the chemicals that surround them. As Deneubourg and Goss (1989) showed, the process by which excited ants infect their immediate neighbors can also be considered autocatalytic.

A big difference between slime molds and ants is that the amoebae are always bounded by a sheath, while the ants roam free. Even if the amoebae simply adhered to one another to varying degrees, it would make them very different from the free-moving social insects.

Another important comparison involves pattern formation. The proportions of different ant or termite castes are regulated by pheromones that control the development of juveniles, so that the ratios of the different castes remain constant. These chemical messages, distributed from one individual to another, are often inhibitors that repress the development of a particular caste. Previously I mentioned the work of Light (1942–1943) on termites; he showed that if the soldiers were removed from a colony, juveniles would, at the next molt, metamorphose into soldiers until their original ratio in the colony was restored. However, if those juveniles were fed a paste made of dead soldiers no such metamorphosis occurred; the inhibitor in the paste prevented the switch in the developmental pathway.

Since insects are free-moving and not bound together in a fixed way, their pattern is quite different from that of slime molds and embryos in general. However, Okamoto (1981) showed that slime mold amoebae in liquid culture, where the cells are floating free and quite separate from one another, can nevertheless regulate the proportions between prestalk and prespore cells. It would be interesting to know if there are any differences in the morphogen control of proportions found in cell suspensions compared to what is found in normal slugs.

### *Evolution beyond the Beginning of Multicellularity*

Since I am discussing the biology of slime mold development, it would be appropriate also to consider some aspects of how slime molds evolved after their origin. I have repeatedly pointed out here that the small, simple forms that exist today have been on the earth for millions of years, constantly under the supervision of natural selection, and that this has resulted in many changes, most of which seem to be ones that reinforce the stability of the primitive multicellular development and involve duplications and modulations of the signal-response system. But there are other changes besides, which are further testimony to the extraordinary, far-reaching effects of natural selection on the cellular slime molds.

There are two obvious advantages to the size increase gained by multicellularity in these organisms. One is protection from predators, as has been ingeniously demonstrated by Kessin and co-workers (1996). They added soil nematodes to their slime mold cultures and found that while the worms feasted greedily on the amoebae, they were unable to eat the aggregated cell masses; they were protected by their size and the slime sheath covering them. Furthermore, they showed that the nematodes did swallow the spores, but they passed through the gut undigested, in this way using the feeding nematodes as a means of dispersal.

As with many fungi, spore dispersal appears to be of great importance for reproductive success of slime molds. So far I have said only that sticking up into the air is the key—in this way the spores will more effectively spread. There is reason to suspect that near the soil surface and in the humus where these organisms live, passing animals such as mites, worms of various kinds, and other motile invertebrates that brush against the spore mass pick up the sticky spores and carry them to some virgin patch of bacterial food. Now I want to show that these simple bags of aggregated amoebae do much more than just stick their spores up into the air to be snagged by passing beasts: they go to quite extraordinary lengths to see that the spores end up in the optimal place for dispersal. I will not give all the experimental evidence here for the basis of this assertion; I will only give the results and describe the skills of the aggregated slime-mold cell mass.

It would appear that most of the slime molds' abilities center around getting the spores from the deeper moist feeding area nearer to the surface of the soil, where apparently there is a better chance of catching a ride. The migrating cell masses are extremely sensitive to light and will go towards light of surprisingly low intensities. Since daylight or moonlight will always come from above, this is one powerful way to orient them upward in the soil.

There is more. Slime molds are also highly sensitive to heat gradients and will orient in gradients so shallow that the temperature difference between the two sides of a small slug need only be 0.0005 °C (Bonner et al., 1950) in order for them to go to the warmer side. We first thought orientation only occurred towards warmer temperatures, but Whitaker and Poff (1980) showed that this was only true if the slugs are migrating in a temperature range above that in which they had been raised; if the gradient was in a colder range they would go towards the colder side, that is, they are negatively thermotactic. Whitaker and Poff gave a neat and convincing explanation to this reversal of orientation: in daytime the sun would make the air and the surface of the soil generally warm so the migrating slugs would crawl towards the warm surface, but at night the heat gradient is reversed and the soil will be warmer than the cool night air; yet since in a cooler environment they are negatively thermotactic they will still go upward. By this intriguing mechanism they orient correctly towards the surface both night and day.

It is also known that cellular slime molds orient by exuding a gas which not only orients slugs, but more importantly positions the rising fruiting body. This gas, which we now know is ammonia, repels the leading tip, so if two fruiting bodies arise close to each other they will lean away from each other. This is because the ammonia that the developing fruiting bodies give off is more concentrated between them, and the sensitive tip will point away, causing the rising cell masses to move away from each other. One can also show that they not only avoid other fruiting bodies but for the same reason they will move away from a wall—in fact, it is this mechanism which makes them rise at right angles from the substratum. If they are in a small cavity in the soil they will, by this gas orientation, position themselves in the dead center of the cavity, and all these sensitive orientations contribute to putting the spores in the ideal place for dispersal. These are remarkable feats for a bag of amoebae.

From all this we see that during the long course of evolution the numerous experiments in becoming multicellular were really just small steps compared to the wonders that followed.

## The Molecular Approach

Since the middle of the twentieth century there has been a great wave of activity on the molecular development of many organisms, including slime molds. Its current success is the direct result of the extraordinary advances in manipulative molecular genetics. It began with the development of *Drosophila* when Christine Nüsslein-Volhard and Eric Weischaus opened the door by finding a way to identify genes and their proteins involved in early embryonic development. As a result of their pioneering work, the study of the development of *Drosophila* is now a major industry, with many laboratories involved, and the progress has been particularly exciting and rewarding. Through the efforts of Sydney Brenner a similar approach was subsequently applied to an organism that has a very different development, the nematode *Caenorhabditis*, and again both the number of workers and the progress have been enormous. Another example is the flowering plant *Aribidopsis*, where large numbers of the genes and proteins related to development have been, and are continuing to be, identified, giving a totally new lease on the study of higher plant development. Now, as never before, it has been possible to piece together development of organisms so that the fine detail of the steps can be unveiled in a way that was inconceivable in the first half of the twentieth century. Furthermore, because of this variety of different organisms, and consequently the variety of radically different developments, it is now possible to do comparative molecular developmental biology, hitherto no more than an impossible dream. There are other organisms that could be added to the list, including vertebrates such as the zebra fish, and it is of particular interest to add cellular slime molds because their development is so radically different.

In studying the molecular development of any organism, we are not just concerned with the signaling between cells, which is the emphasis here, but within cells as well. After all, the whole point of modern molecular developmental biology is ultimately to understand how the genes direct development. So we begin in the genome and follow the sequence through protein synthesis and the developmental activities of those proteins. It is a complex subject for many reasons. There is the problem of how the timing of a particular developmental event is controlled, and we must determine how the genes control pattern. Progress in the field in general has been spectacular, but the success has brought new problems.

To begin, we find that there are often many genes that control a single step. Those sets of genes work in harmony, so that some activate, and some inhibit, and some simply modulate the ultimate effect. In turn, the gene-produced proteins are very busy in their actions and interactions within the cell. They may be enzymes that are responsible for the synthesis of other molecules that play a key role in development; they may be structural proteins that are directly responsible for the shape or the locomotion of the cell, and there are a myriad of other important biochemical pathways. The point I want to make here is that all those gene products, and the descendants of those products, interact with one another inside the cells—it is not just the genes that affect one another, but so do their products, which in turn may affect the genes. There is a great interacting net of genes and products that steers the course of development.

One of the successes of this molecular revolution has been in the evolution of development. There are certain developmental genes that can be traced in different groups of animals where they serve the same or different functions. The best-known example is the cluster of Hox genes in animals, where their sequence on the chromosome is reflected in the sequence in the segmented anatomy they are responsible for creating. Their lineage can be traced from invertebrates through mammals, where both the genetic changes and the morphological changes can be followed. It is a fascinating story in all its details.

What I am going to do here is discuss the molecular aspects of cellular slime development. It will be a most cursory and incomplete review (see Kessin, 2000, for a serious and comprehensive one). In fact, I will concentrate only on some of those aspects that have already been introduced in the earlier biology section of this chapter. Since the stress here has been on signaling between cells, one of the things that will be obvious is that those signals emanate from within the cells with all the complications I have just alluded to, and turn into a comparatively simple communication system once we get outside the cell membrane.

### *The Study of Slime Mold Genes and Their Activities*

Let me briefly indicate how one goes about looking at the molecular side of development. Ultimately it is based on genetics. Unfortunately, it has not been possible to exploit the sexual cycle of cellular slime molds to make crosses, but by various less direct methods the known genes can be assigned into linkage groups. There are six chromosomes or linkage groups, and for comparison, the entire genome is only one percent of the size of the human genome (although three times greater than yeast, and eight times greater than the bacterium *E. coli*). A number of research groups are presently cooperating to sequence the entire genome of *Dictyostelium discoideum,* and obviously that will allow for rapid advances in uncovering the genes specifically involved in development.

At the moment, the principal ways one can study the all-important question of what particular genes are doing in development is by modifying their expression in some way. This can be done by deleting the gene (null mutation); or knocking out the gene by inserting a blasticidin-resistant gene in its place and selecting for it in the presence of blastocidin; or by silencing the gene with RNA binding to a specific region of DNA. One can even overexpress a gene by inserting a more powerful promoter by it. Finally, it is possible to insert new genes. These ways of changing a gene or inserting new genes are all part of the clever armamentarium of modern molecular biology. These tricks can be done to individual amoebae, which are then cloned. There is, of course, much more to it than these few words suggest, but this sort of manipulation is the main way to understand what genes are doing in development.

Another technique now widely used for developmental studies in a wide variety of organisms is to insert the gene for the Green Fluorescent Protein (GFP) by the promoter of a gene for a protein that is of developmental interest. With such a marker one can, in slime molds, follow various proteins associated with stalk and spore differentiation; it is a wonderful tool for investigating the microanatomy of cell fates and the distribution of different cell types during morphogenesis.

Then, besides the genetics, there is always a lot of straight biochemistry that goes with it. Let us now look at our previously mentioned topics of polarity, chemotaxis, and pattern formation from these more molecular points of view.

### *Polarity*

As pointed out in the previous discussion, polarity is an alignment, the onset of a direction to the embryo or the cell mass. It is often something triggered by an environmental cue, and once established it in turn provides information for the establishment of pattern. From the work on slime mold polarity, it is clear that oxygen is one of the cues that can establish its polarity, and there are two interesting features about the way this can happen. One is that the cell masses are sensitive to exceedingly small differences in oxygen tension, and the other is that oxygen can exert its influence rapidly—in a minute's time in the capillary experiments.

Here we have an instance where we know nothing of the biochemistry and certainly nothing of the molecular biology. In fact, the latter is ruled out automatically because one minute is far too little time for gene transcription and translation to occur. And no one has even looked at the biochemistry, something that will be very interesting for the future. What are the substances that respond so rapidly to small differences in oxygen concentration, and how do they transfer that signal into rapid amoeboid motion to the cells? Clearly, whatever those substances are, it will eventually be possible to trace their origin to gene products and to the genes that produced those products, but the most interesting aspect is the mechanism of the series of rapid events. There are no doubt many other bits of important biochemistry in slime mold development—as well as in the development of other organisms—of which we know nothing, and perhaps there are others of equal interest and importance whose very existence is still unknown. There is a bright future for this approach to development.

### *Chemotaxis*

Since the early days we have known that slime mold aggregation is by chemotaxis, and since the 1960s we have known that the attractant for many species is cAMP. It was shown early that there is a relay, and that if a cell is hit with a puff of cAMP at its anterior end (or if it is in a gradient of the attractant), the cell will move towards the high concentration of the cAMP and rapidly release from inside the amoeba more cAMP into the outside. This rapid cAMP stimulus-response mechanism is immediately followed by a refractory period where the cell, for an interval, is incapable of responding to cAMP. Note that this is another way of imparting polarity—in this case to the individual amoebae.

All this is essentially biological information that generated great interest in examining the biochemical basis of chemotaxis, a study that continues to be very active today. For instance, there have been successful attacks on the cAMP receptors on the cell surface and an understanding of how the activated receptor passes the message along, a complex process that involves a number of associated proteins. A great deal is known about such transduction pathways in other organisms, providing an opportunity for some interesting comparative biochemistry. In chemotaxis, it turns out that the cAMP receptors themselves stay distributed over the surface of the amoeba, while others of these transduction pathway proteins accumulate at the anterior end of the cell (review: Parent et al., 1998).

Chemotaxis might have initially arisen very early in the invention of aggregation organisms. What is striking is the number of genes and their proteins that are involved in slime mold chemotaxis. The ancestral amoebae were possibly as internally complex then as they are now; the interesting question is how did they use all the cellular machinery to communicate between cells. No doubt they put molecules that were already there, such as cAMP, to some new or additional use, in this case a chemoattractant.

### *Adhesion Molecules and the Slime Sheath*

Another very important feature of slime mold togetherness, alongside chemotaxis, is cell adhesion and the secretion of a slime sheath which covers all the cells in the mass. Much has been done on this interesting subject; specific proteins and glycoproteins are involved, and by isolating and identifying these molecules, and in some cases identifying their genes as well, it has been possible to gain considerable insight into their function. These adhesion and slime molecules are secreted by the cells, and they play a key role in aggregation and all the later stages of development. By using various methods of blocking a gene so that a specific adhesion molecule fails to be produced, one can observe the defect. In this way, it has been possible to show that there are numerous adhesion molecules with discrete functions. If the appropriate genes are blocked, the cells will aggregate into a loose rubble that is incapable of further development. Clearly, these molecules are essential for normal development, not only to hold the cells together, but also, no doubt, to signal between the cells. (For a review, see Kessin, 2000.)

### *Control of the Stages of Development*

There has been a sustained interest and continual progress in our understanding of the signals that push the amoebae through the various stages of their life cycle. The transitions are as follows: the germination of the dormant spore;

the jump from feeding stage to the initiation of aggregation; the formation of a tip at the mound; the initiation of prestalk and prespore differentiation in the migrating slug; the transition from migration to the initiation of culmination or fruiting; and the formation of final spore and stalk cell differentiation. We know nothing of the molecular basis for some of these transitions, but for others there has been good progress.

We know quite a bit of the details of spore germination; we know that the initiation of aggregation is begun by cells that began starving just before they were about to divide—they are the amoebae that first start giving off cAMP signals (McDonald, 1986); we know that the shift from migration to culmination is under the control of ammonia, and high ammonia means continued migration (Schindler and Sussman, 1977); and in the discussion below we will see that a lot is also known about the molecular events leading to final stalk cell and spore differentiation. Nevertheless, we still have many gaps in our knowledge.

### *Pattern Formation*

To modern developmental biologists, pattern remains the issue of central interest, as indeed it has been since the middle of the nineteenth century. In the case of cellular slime molds, the key question is: What are the biochemical control mechanisms that ordain a fixed proportion of stalk and spore cells? These studies have concentrated on *D. discoideum* because not only are the proportions retained in the stalk and spore cells in the final fruiting bodies, but, as we saw, so are the prestalk and prespore zones in the migrating slugs.

We are very far from having a satisfactory picture of what transpires. We do know that it is possible to influence the likelihood that a cell will become either stalk or spore in a number of ways: cells reared on a rich medium, when mixed with ones fed on a minimal medium, will tend to become spores, while the ones raised on the thinner diet become stalk cells (Leach et al., 1973); cells that are about to divide become spores (i.e., they are replete with food), while cells that have just divided tend to become stalk cells (McDonald and Durston 1984). The explanation of how such predisposed cells get to the right place has to do with the ability of cells to sort out within the aggregate and with early slug formation (see, for instance, Bonner, 1959; Takeuchi and Sato, 1965; Sternfeld and David, 1981a).

However, it is quite certain that this is not the mechanism that produces the correct ratio of roughly 20 percent stalk cells and 80 percent spores. This ratio is to some degree affected by the size of the cell mass (see Nanjundiah and Bhogle, 1995, for details). As Raper (1940) demonstrated many years ago, if a portion of the slug consisting entirely of cells that were destined to become either stalk or spores is isolated, it will eventually regulate and again form a normal, small slug and later a fruiting body with the usual ratio of stalk cells and spores.

The important message is that even though we have many factors that favor cells leaning in either the stalk or in the spore direction—such as how replete the cells are with food and where they lie in an oxygen gradient—in the end there is clearly some chemical communication between the anterior and the posterior cells which establishes the sharp division line between the number of stalk cells and the number of spores. This is the big question that has been attacked in a number of laboratories. Without going into the details, let me say that there are numerous diffusible substances that appear to play a role. These include cAMP, ammonia, DIF (Differentiation Inducing Factor, a small molecule discovered by R. Kay that specifically induces stalk cells), adenosine, and calcium (for reviews, see Nanjundiah and Saran, 1992; Gross, 1994; Kessin, 2000). If one reads about what is known of the patterning action of these morphogens, two things are quite clear: the relationship between them has not been unscrambled, but when that happens, there will be more substances involved of which so far we only have inklings. Some of the morphogens are activators, some are inhibitors, and by a sort of chemical conversation among them they manage to fine-tune the division line into the right place, even when fractions of a slug are isolated. Remember that the anterior cells are churning about in a chaotic fashion, as we saw in the two-dimensional slugs. These morphogens must therefore act primarily extracellularly, but their actions take place within the slime sheath. Furthermore, it is now known through some clever experiments labeling specific proteins that the prestalk region is actually a composite of discrete subzones (Williams, 1997).

### *Conclusion*

The study of the molecular development of cellular slime molds is giving us a deep insight into how development is controlled, how it is directed on the microlevel. The science of molecular biology has made it possible to dissect at ever lower levels of analysis. We are trying to understand causes at the most reductionist level. It is the dream that started in the nineteenth century with Wilhelm Roux's *Entwicklungsmechanik*, his mechanics of development. Today we can go back one large step further in the chain of causes that brings us from egg to adult, or in the case of our slime molds, from spore to adult. This is clearly not only desirable, but an essential endeavor for modern biology. It is like following a rainbow at breakneck speed looking for the pot of gold at the end.

My message here is somewhat different from the standard message of molecular biology. I fear that the search for the gold may involve an infinity of regressions. Perhaps this is not so, and certainly I wish it were not, but I feel there must be some other way of looking at the problem that will give another kind of satisfaction—perhaps a more immediate one. As I have argued, part of the reason for the plethora of morphogens in slime molds no doubt stems from the fact that they have been subject to change by natural selection for many millions of years. As a result, the essence of development has been obscured by the great accumulation of refinements over this vast amount of time. We cannot go back to their primeval ancestors to find out the initial morphogen signaling system that they first adopted on the threshold of multicellularity. All we can do is ask what it might have been—what is the simplest way involving the fewest morphogens to account for slime mold development. To answer this question, we must turn to mathematical modeling.

## The Mathematical Modeling Approach

My use of modeling will be different from the usual ones. I will not look for models that make testable predictions, and I will not look for models that can be ever modified and expanded so that they can encompass all the details of a real phenomenon as the empirical details come to light. Rather, I will turn to models that illuminate the simplest solution.

There is a great interest in mathematical biology and the use of models in many disciplines within biology, such as ecology, evolutionary biology, physiology, cell biology, and developmental biology. The important question is what has mathematics done, or what can it do for biology. That is not always an easy question to answer.

The most obvious way of looking at the relationship is to point out that biology, in all its parts, invariably involves an enormous amount of detail: animal and plant diversity, complex interactions between organisms in any one environment, the vast molecular activities within a single cell, the size of the genome, the complexities of animal behavior, and so forth. One way in which mathematics has been a been a clear success is in organizing such tangles by turning what seems to be chaos into some kind of order. The worry is always: Does the simplification distort reality, and are all the simplifying assumptions reasonable?

Often one criterion for suitability of a model is to see if it makes predictions that can be tested. If the predictions are supported, it pleases both the modeler and the empiricist. Some of the mathematical models in ecology, especially those put forward in early days, were a revelation because they showed that there was some kind of sense, of meaning, in what previously had seemed a jumble of observations. Predictions can often be made in ecological models; those predictions can be tested and the model changed so that it will come closer to reality. From the very beginning it was appreciated that the models were approximations and not reality, and that every model might be the precursor of a better one that fits more closely what is found in nature.

In developmental biology, the disparity between the models and what one finds in embryos has been large. The main reason is the one that has been repeatedly mentioned here: there is a great multiplicity of signaling morphogens and their corresponding receptors in development. This is the reality, and this is precisely what the models ignore, and the simplifications are often considered misleading and naive. Even when a great effort is made to match reality, new experiments and empirical detective work produce additional details of the signaling complexes that must accordingly be further accommodated.

If one compares developmental biology and ecology, no doubt one finds a cultural difference between the two disciplines, but more important is the difference in their very nature. In a developing embryo one can see the changes as they occur, and by experiment it has been possible, beginning way back into the nineteenth century, to examine on a crude level how one change leads to the next. Embryos do not begin as a confusing jumble of morphogens, but they are concrete objects that go through a series of changing morphological (and biochemical) steps, where each step leads to the next in a causal sequence. Now we know much more, and each of those steps has become increasingly complex; what we have discovered below the surface is a thicket of actions and reactions. How can we possibly say what the first steps were like in early earth history, those that would give us a clue to the basic, underlying mechanism of development?

The modeler asks what the simplest explanation for any phenomenon is, and then expresses that simplicity in the least elaborate set of equations. What he or she is saying is that the actual biological process may be far more complex, but in development this particular form, or change in form, need only consist of the following few elements. The approach, then, by its fundamental nature, is perfectly suited to suggest some ideas of what might have occurred at the beginning of multicellular development. Obviously, it cannot say that this is indeed what happened, but it does give us a way of thinking about that beginning which in itself is a great boon. The modern-day complexities of development are a reality, but we are asking, how did it begin? So we are not looking for models that make predictions that can be tested, but ones that can give us an idea of the kind of thing that might have happened initially.

### *Hypothetical First Steps in Slime Mold Development*

Before we think of mathematical models, it will be helpful to ask what might have been the minimal sequence of initial steps in the rise of slime mold multicellular development. My plan is to begin with a nonmathematical model.

The obvious presumption is that the following sequence is the direct result of natural selection: (1) The first of such steps would be that an aggregation of spores or cysts might, in some ecological circumstances in the soil, have an advantage over solitary ones. (2) Again, under some circumstances, it might be advantageous for the collection of resistant spores to be lifted into the air. (3) The most effective way to achieve the latter is to have some of the cells in some manner sacrifice themselves to build a stalk, while the remaining cells turn into spores. (4) In order for this process to be effective, it must be consistent each generation and for different sizes of cell masses.

I will stop with these four steps and not consider any of the refinements that follow, such as the ability to regulate after the cutting of the cell mass, or the clever devices that have evolved to see that fruiting occurs in the optimal place for dispersal, and other advantageous processes that might have been further encouraged by natural selection.

Aggregation is, from the beginning, a biochemical problem. Amoebae are motile and a stimulus-response mechanism must be devised so that they respond by moving up a gradient of the signal molecule. I have already made the point that for this, some small, diffusible molecule that might serve other functions within the cells could be appropriated, and gave cAMP as the example.

We are left with the problem of pattern: How can the right number of prestalk cells occupy the anterior 20% of the slug and the prespore cells form 80% of the posterior end? This is a question that can be helped by a mathematical model.

## *Turing's Reaction-Diffusion Model*

There are many possible models for pattern in biological form and many excellent books on the subject. (To mention a few: Meihnardt, 1982; Segel, 1984; Murray, 1989; Ball, 1999; see Cox, 1992, for an interesting minireview.) One of the most successful approaches stems from the famous paper of Turing (1952) and involves both the diffusion of key molecules and their chemical reactions. To give one example of many, Nijhout (1991) has made a detailed and successful exploration of the patterns on butterfly wings in terms of reaction-diffusion equations. There are also many applications of the Turing approach to slime molds that are elegantly summarized by Nanjundiah (1997).

As will become evident, my intention is not to repeat what has been so thoroughly done before, namely to modify the basic Turing model to meet all the details of a specific case, but rather to examine his model in its minimal form to see if it might give us some helpful insights into origins.

Alan Turing's idea is simplicity itself. The requirements for his pattern-making machine are two basic substances. One is an activator molecule that, among other things, can activate itself—it is autocatalytic—and it can activate other cellular processes. The other is an inhibitor that counters the effects of the activator by repressing the actions of the activator. The only further requirement is that something is needed to set the initial path of the diffusion of these substances and their subsequent reactions; in other words, they need to be provided with a polarity.

It might be helpful to illustrate the basic idea with the work of Gierer and Meinhardt (1972) on the freshwater *Hydra*. They assumed a polarity to the individual, and as a result of this polarity the two diffusing substances produced are highest at the anterior end. One of those substances, the activator, has the property of being autocatalytic; that is, it stimulates its own production and therefore steadily increases in concentration. The other is an inhibitor that specifically inhibits the production of the activator. The other important requirement is that the inhibitor be a smaller molecule and diffuse faster than the larger activator. Under these rules it is possible to predict the concentration of both substances along the axis of the *Hydra*: the activator will be highest at the anterior end, while the inhibitor will be highest at the posterior end, a matter that is best illustrated in a simple graph (fig. 30).

### *Cellular Slime Molds*

It is an easy matter to apply these principles to slime molds. As was described earlier, the migrating slugs have, regardless of their size, consistent proportions between the anterior prestalk and the posterior prespore zones, and this is readily modeled as a reaction-diffusion pattern (Meinhardt, 1982; MacWilliams and Bonner, 1979; see Nanjundiah, 1997, for the details of the different models). The point here is that even though we know that a number of morphogens are involved in slime molds that live today, the model tells us that the same thing could be accomplished with merely one activator and one inhibitor. These could be cAMP, which is known to stimulate its own production, and ammonia, which is known to inhibit the pro-

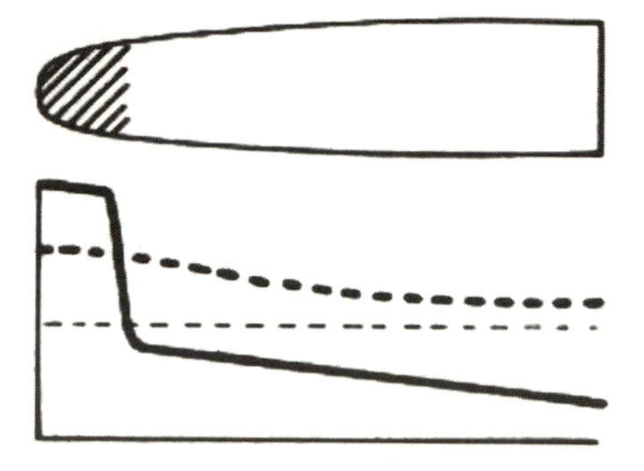

Fig. 30. A graph illustrating how pattern can be generated by a reaction-diffusion system, in this case the sharp delineation of the prestalk and prespore zones in a slime mold migrating slug. The concentration of both the activator (solid line) and the inhibitor (thick dashed line) increase in the front end of a slug, and once the activator rises above a threshold (thin dashed line) the prestalk-prespore division line is formed. (From MacWilliams and Bonner, 1979.)

duction of cAMP. This is, of course, all pure speculation—unfortunately, no one was there to admire the origin of the ancestor of all the living cellular slime molds. However, the model gives us an insight as to how it all could have started. We will never know what really happened so many years ago, but by making mathematical models we can see that the beginning could indeed have been simple, and that simplicity underlies all the complications we see on the earth today.

# 8 *Conclusion*

HERE ARE THE bare bones of my argument. A basic premise from which all else follows is that most, if not all, the microorganisms that exist today have an ancestry that goes back many millions, if not billions of years. On this foundation I have argued that (1) the size of organisms is under constant selection pressure, and in early earth history, when all of life was unicellular, one easy and frequently occurring way of becoming larger was to become multicellular; (2) as soon as such cell conglomerates appeared, there was selection for integration between the cells, and this involved the production of extracellular signaling systems that led to cell differentiation and pattern, all built on the foundation of environmentally induced polarity; (3) because there immediately followed a nonstop selection for stabilizing and modulating these signal-response systems, they became more elaborate, more complex, so that in the development of the primitive organisms we study today the initial simplicity is buried in a great mass of additional pathways, and therefore is no longer possible to see. In trying to reconstruct the beginning, our most effective tool is mathematical modeling, which allows us to ask: What is the minimum signaling needed to produce a pattern? I will now discuss and review these three points.

# Size

Presumably the first multicellular groups of cells were accidents; perhaps they were the result of the production of a new adhesion molecule that kept the cells glued together. Because of this fortuitous increase in size, certain advantages became the automatic consequence. The example I gave was the obligate anaerobe *Methanosarcina,* which, by forming a clump, could keep its internal cells shielded from the deadly effect of poisonous oxygen. Protection from the environment could have been a general initial advantage for this first step in size increase. Further advantages would have followed, such as more effective sun catching for photosynthesis, more effective means of dispersal, more rapid locomotion, and perhaps other hidden gains that we have not yet realized. All of these will also have been immediately subject to natural selection, with the result of increased efficiency. Furthermore, there will be a constant open opportunity for larger multicellular forms that could be even more efficient and exist in a competition-free world.

There is an enormous range in the sizes of living organisms—from bacteria to whales and giant sequoias—and it is not surprising that there is every level of intermediate size as well. It is self-evident that the world could not exist with just large animals and trees; there are many reasons for this, with energy requirements being paramount. This means the world of organisms is made up of niches, and the kind of niche that has concerned us here is a size niche.

An important point to stress is that the niche for the smallest organisms is the most ancient. This means that its inhabitants have had not only much more absolute time, but also lived through many more generations because of the combination of their early origin and their short generation time due to their small size. So the first steps are overlaid with a prodigious number of opportunities for change, for increases in complexity.

## INTEGRATION

Selection for efficiency in the first cell masses meant integration, which in turn meant extracellular signals and responses—in other words, communication between the cells. One of the prime roads to efficiency was through cell differentiation, that is, a division of labor among the cells.

However, it must have been true from the very beginning that the environment provided what were undoubtedly the first signals. In a crude way, this is what we assumed happened in the case just mentioned of *Methanosarcina.* The main point is that the polarity of the cell mass is likely initially to have been imparted directly by the environment. This may have occurred by gradients of a gas, such as oxygen, or a physical entity such as directional light. It is interesting that all living organisms today still begin their development from some physical or chemical environmental cue that imparts their polarity. In some ways, the establishment of polarity in animals and plants today is a kind of living developmental fossil.

## Mathematical Models

As to the beginning of cell-cell signaling, we are pretty much in the dark because there have been so many overlays of additional regulatory signals over the vast amount of intervening time. The best recourse for dealing with this difficulty is to see what we can do by mathematical modeling, and here the point was illustrated by applying Turing's reaction-diffusion model to the origin of cellular slime mold development.

There is a great conflict with modeling. So much of it involves trying to marry empirical data with the best-fitting equations, with a great emphasis on testing the predictions of the model. During the process it is often possible, and very gratifying, to gain insights and order from a confusing set of descriptive facts. But here the models are used entirely for fantasyland; it is a way of asking how it might have been in the beginning. It is our greatest tool to ask that question, free of details and frills. It can tell us what is at the very foundation of all of development.

# BIBLIOGRAPHY

Baker, J. R. 1948. The status of the protozoa. *Nature* 161: 548–551, 587–589.

Baldauf, S. L. 1999. A search for the origins of animals and fungi: Comparing and combining molecular data. In *Evolutionary Relationships among Eukaryotes*, ed. L. A. Katz. *Amer. Naturalist* 154 (suppl.): S178–S188.

Ball, P. 1999. *The Self-Made Tapestry: Pattern Formation in Nature.* Oxford: Oxford University Press.

Bates, H. W. 1863. *The Naturalist on the River Amazons.* London: John Murray.

Beadle, L. C., and F. A. Booth. 1938. The reorganization of tissue masses of *Cordylophora lacustris* and the effect of oral cone grafts, with supplementary observations on *Obelia gelatinosa. J. Exper. Biol.* 15: 303–326.

Bell, G. 1985. The origin and early evolution of germ cells as illustrated in the volvocales. In *The Origin and Evolution of Sex*, ed. H. Halvorson and A. Monroy, 221–256. New York: Alan R. Liss.

Birch, C. L., and L. Chao. 1999. Evolution by small steps and rugged landscapes in the RNA virus ø6. *Genetics* 151: 921–927.

Bonner, J. T. 1949. The demonstration of acrasin in the later stages of the development of the slime mold *Dictyostelium discoideum. J. Exp. Zool.* 110: 259–271.

Bonner, J. T. 1952a. *Morphogenesis: An Essay on Development.* Princeton: Princeton University Press.

Bonner, J. T. 1952b. The pattern of differentiation in amoeboid slime molds. *Amer. Naturalist* 86: 79–89.

Bonner, J. T. 1958. *The Evolution of Development.* Cambridge and New York: Cambridge University Press.

Bonner, J. T. 1959. Evidence for the sorting out of cells in the development of the cellular slime mold. *Proc. Natl. Acad. Sci. USA* 45: 379–384.

Bonner, J. T. 1965. *Size and Cycle.* Princeton: Princeton University Press.

Bonner, J. T. 1967. *The Cellular Slime Molds.* 2nd ed. Princeton: Princeton University Press.

Bonner, J. T. 1974. *On Development: The Biology of Form.* Cambridge, Mass.: Harvard University Press.

Bonner, J. T. 1988. *The Evolution of Complexity.* Princeton: Princeton University Press.

Bonner, J. T. 1996. *Sixty Years of Biology: Essays on Evolution and Development.* Princeton: Princeton University Press.

Bonner, J. T., and M. R. Dodd. 1962. Aggregation territories in the cellular slime molds. *Biol. Bull.* 122: 13–24.

Bonner, J. T., W. W. Clarke, Jr., C. L. Neely, Jr., and M. K. Slifkin. 1950. The orientation to light and the extremely sensitive orientation to temperature gradients in the slime mold Dictyostelium discoideum. *J. Cell. Compar. Physiol.* 36: 149–158.

Bonner, J. T., K. B. Compton, E. C. Cox, P. Fey, and K. Y. Gregg. 1995. Development in one dimension: The rapid differentiation of *Dictyostelium discoideum* in glass capillaries. *Proc. Natl. Acad. Sci. USA* 92: 8249–8253.

Bonner, J. T., P. Fey, and E. C. Cox. 1999. Expression of prestalk and prespore proteins in minute, two-dimensional *Dictyostelium* slugs. *Mechanisms of Devel.* 88: 253–254.

Bonner, J. T., L. Segel, and E. C. Cox. 1998. Oxygen and differentiation in *Dictyostelium discoideum. J. Biosci.* 23: 177–184.

Buss, L. W. 1982. Somatic cell parisitism and the evolution of somatic tissue compatibility. *Proc. Natl. Acad. Sci USA* 79: 5337–5341.

Buss, L. W. 1987. *The Evolution of Individuality.* Princeton: Princeton University Press.

Buss, L. W. 1999. Slime molds, ascidians, and the utility of evolutionary theory. *Proc. Natl. Acad. Sci. USA* 96: 8801–8803.

Child, C. M. 1941. *Patterns and Problems of Development.* Chicago: University of Chicago Press.

Cox, E. C. 1992. Modeling and experiment in developmental biology. *Current Opinion in Genet. and Develop.* 2: 647–650.

Dawkins, R. 1976. *The Selfish Gene.* New York: Oxford University Press.

Deneubourg, J. L., and S. Goss. 1989. Collective patterns and decision making. *Ethol. Ecol. and Evol.* 1: 295–311.

Driesch, H. 1908. *The Science and Philosophy of the Organism.* London: A. and C. Black.

Dworkin, M. 1972. The myxobacteria: New directions in studies of prokaryotic development. *Crit. Rev. in Microbiol.* 2: 435–452.

Dworkin, M., and D. Kaiser, eds. 1993. *Myxobacteria II.* Washington, D.C.: American Society for Microbiology.

Dyson, F. 1999. *Origins of Life.* New York: Cambridge University Press.

Filosa, M. F. 1962. Heterocytosis in cellular slime molds. *Amer. Naturalist* 96: 79–92.

Gerhart, J., and M. Kirschner. 1997. *Cells, Embryos, and Evolution: Toward a Cellular and Developmental Understanding of Phenotypic Variation and Evolutionary Adaptability.* Oxford: Blackwell Scientific.

Gierer, A., and H. Meinhardt. 1972. A theory of pattern formation. *Kybernetik* 12: 30–39.

Goebel, K. 1900–1905. Organography of plants, especially of the Archegoniatae and Spermaphyta. Oxford: Clarendon Press.

Gross, J. D. 1994. Developmental decisions in *Dictyostelium discoideum. Microbiol. Rev.* 58: 330–351.

Gurdon, J. B., S. Dyson, and D. St. Johnston. 1998. Cell's perception of position in a concentration gradient. *Cell* 95: 159–162.

Hamburger, V. 1988. *The Heritage of Experimental Embryology: Hans Spemann and the Organizer.* Oxford and New York: Oxford University Press.

Haselkorn, R. 1998. How cyanobacteria count to 10. *Science* 282: 891–892.

Hutchinson, G. E. 1959. Homage to Santa Rosalia, or why are there so many kinds of animals? *Amer. Naturalist* 93: 145–159.

Huxley, J. S., and G. de Beer. 1934. *Elements of Experimental Embryology.* Cambridge and New York: Cambridge University Press.

Hyman, L. 1940. *The Invertebrates: Protozoa through Ctenophora.* New York: McGraw-Hill.

Jaffe, L. 1968. Localization in the developing *Fucus* egg and the general role of localizing currents. *Adv. Morphog.* 7: 295–328.

Kessin, R. 2000. *Dictyostelium: Evolution, Cell Biology and the Development of Multicellularity.* Oxford and New York: Oxford University Press.

Kessin, R. H., G. G. Gundersen, V. Zaydfudim, M. Grimson, and R. L. Blanton. 1996. How cellular slime molds evade nematodes. *Proc. Natl. Acad. Sci. USA* 93: 4857–4861.

Kirk, D. 1998. *Volvox: Molecular-Genetic Origins of Multicellularity and Cellular Differentiation.* Cambridge and New York: Cambridge University Press.

Kosugi, T., and K. Inouye. 1989. Negative chemotaxis to ammonia and other weak bases by migrating slugs of the cellular slime moulds. *J. Gen. Microbiol.* 135: 1589–1598.

Leach, C. K., J. M. Ashworth, and D. R. Garrod. 1973. Cell sorting out during the differentiation of mixtures of metabolically distinct populations of *Dictyostelium discoideum. J. Embryol. Exp. Morphol.* 29: 647–661.

Light, S. T. 1942–1943. The determination of the castes of social insects. *Quart. Rev. Biol.* 17: 312–326; 18: 46–63.

McDonald, S. A. 1986. Cell-cycle regulation of center initiation in *Dictyostelium discoideum. Develop. Biol.* 117: 546–549.

McDonald, S. A., and A. J. Durston. 1984. The cell cycle and sorting behaviour in *Dictyostelium discoideum*. *J. Cell Sci.* 66: 195–204.

MacWilliams, H. K., and J. T. Bonner. 1979. The prestalk-prespore pattern in cellular slime molds. *Differentiation* 14: 1–22.

Maynard Smith, J., and E. Szathmary. 1995. *The Major Transitions in Evolution*. San Francisco: Freeman.

Meinhardt, H. 1982. *Models of Biological Pattern Formation*. New York: Academic Press.

Miller, J. 1937. Some effects of oxygen on polarity in *Tubularia*. *Biol. Bull.* 73: 369.

Michod, R. E. 1999. *Darwinian Dynamics: Evolutionary Transitions in Fitness and Individuality*. Princeton: Princeton University Press.

Morgan, T. H. 1903. Some factors in the regeneration of *Tubularia*. *Roux Arch. Entwickl.* 16: 125–154.

Moser, J. C., R. C. Brownlee, and R. Silverstein. 1968. *J. Insect Physiol.* 14: 529–535.

Murray, J. D. 1989. *Mathematical Biology*. New York: Springer-Verlag.

Nanjundiah, V. 1997. Models for pattern formation in the Dictyostelid slime molds. In *Dictyostelium—A Model System for Cell and Developmental Biology*, ed. Y. Maeda, K. Inouye, and I. Takeuchi, 305–322. Tokyo: Universal Academy Press.

Nanjundiah, V., and A. S. Bhogle. 1995. The precision of regulation in *Dictyostelium discoideum*: Implications for cell-type proportioning in the absence of spatial pattern. *Indian J. Biochem. Biophys.* 32: 404–416.

Nanjundiah, V., and S. Saran. 1992. The determination of spatial pattern in *Dictyostelium discoideum*. *J. Biosci.* 17: 353–394.

Nijhout, H. F. 1991. *The Development and Evolution of Butterfly Wing Patterns*. Washington, D.C.: Smithsonian Institution Press.

Okamoto, K. 1981. Differentiation of *Dictyostelium discoideum* cells in suspension culture. *J. Gen. Microbiol.* 127: 301–308.

Olive, L. S., and C. Stoianovitch. 1966. *Schizoplasmodium*, a mycetozoan genus intermediate between *Cavosteilium* and *Protostelium*; a new order of Mycetozoa. *J. Protozool.* 13: 164–171.

Parent, C. A., B. J. Blacklock, W. M. Froehlich, D. B. Murphy, and P. N. Devreotes. 1998. G protein signaling events are activated at the leading edge of chemotactic cells. *Cell* 95: 81–91.

Provine, W. B. 1971. *The Origins of Theoretical Population Genetics.* Chicago: University of Chicago Press.

Quatrano, R. S. 1978. Development of cell polarity. *Ann. Rev. Plant Physiol.* 29: 487–510.

Raper, K. B. 1940. Pseudoplasmodium formation and organization in *Dictyostelium discoideum. J. Elisha Mitchell Sci. Soc.* 56: 241–282.

Ray, D. L., and R. E. Hayes. 1954. *Hartmanella astronyxis*: A new species of free-living amoeba. *J. Morphol.* 95: 159–188.

Sachs, J. 1882. *Text-Book of Botany: Morphological and Physiologcal.* 2nd ed. Oxford: Clarenden Press.

Sawada, Y., Y. Maeda, I. Takeuchi, J. Williams, and Y. Maeda. 1998. Rapid patterning of *Dictyostelium discoideum* cells under confined geometry and its relation to differentiation. *Develop. Growth Differ.* 40: 113–120.

Schindler, J., and M. Sussman. 1977. Ammonia determines the choice of morphogenetic pathways in *Dictyostelium discoideum. J. Mol. Biol.* 116: 161–169.

Schneirla, T. C., and G. Piel. 1948. The army ant. *Scientific American* 187: 16–23.

Schopf, J. W. 1993 . Microfossils of the early Archean apex chert: New evidence of the antiquity of life. *Science* 260: 640–646.

Segel, L. A. 1984. *Modeling Dynamic Phenomena in Molecular and Cellular Biology.* Cambridge and New York: Cambridge University Press.

Shaffer, B. M. 1957. Properties of slime mould amoebae of significance for aggregation. *Quart. J. Microscop. Sci.* 98: 377–392.

Smith, G. M. 1955. *Cryptogamic Botany*. Vol. 1. 2nd ed. New York: McGraw-Hill.

Sogin, M. L. 1991. Early evolution and the origin of the eukaryotes. *Current Opinion in Genet. and Develop.* 1: 457–463.

Stebbins, G. L., Jr. 1950. *Variation and Evolution in Plants.* New York: Columbia University Press.

Sternfeld, J., and C. N. David. 1981a. Cell sorting during pattern formation in *Dictyostelium. Differentiation* 20: 10–21.

Sternfeld, J., and C. N. David. 1981b. Oxygen gradients cause pattern orientation in *Dictyostelium* cell clumps. *J. Cell Sci.* 50: 9–17.

Takeuchi, I., and T. Sato. 1965. Cell differentiation and cell sorting in the development of cellular slime molds. *Jap. J. Exp. Morphol.* 19: 67–70.

Trembley, A. 1744. *Mémoires pour servir à histoire d'un genre de polype d'eau douce, à bras en forme de corne.* Leiden.

Turing, A. M. 1952. The chemical basis of morphogenesis. *Phil. Trans. Roy. Soc. London* B 237: 37–72.

Weiss, P. 1939. *Principles of Development.* New York: Henry Holt.

Went, F. W., and K. V. Thimann. 1937. *Phytohormones.* New York: Macmillan.

West-Eberhard, M. J. 2000. *Innovative Phenotypes: Developmental Plasticity and Evolution.* Oxford and New York: Oxford University Press.

Whitaker, B. D., and K. L. Poff. 1980. Thermal adaptation of thermosensing and negative thermotaxis in *Dictyostelium. Exp. Cell Res.* 128: 87–93.

Wilcox, M., G. Mitchison, and R. J. Smith, 1973. Pattern formation in the blue green alga, *Anabaena.* I. Basic mechanisms. *J. Cell Science* 12: 707–723.

Williams, J. 1997. Prestalk and stalk heterogeneity in *Dictyostelium.* In *Dictyostelium—A Model System for Cell and Developmental Biology*, ed. Y. Maeda, K. Inouye, and I. Takeuchi, 293–304. Tokyo: Universal Academy Press.

Winston, M. L., and K. N. Slessor. 1992. The essence of royalty: Honeybee queen pheromone. *Amer. Sci.* 80: 375–385.

Yamamoto, M. 1977. Some aspects of behavior of the migrating slug of the cellular slime mold *Dictyostelium discoideum. Develop. Growth Differ.* 19: 93–102.

Yoon, H.-S., and J. W. Golden. 1998. Heterocyst pattern formation controlled by a diffusible peptide. *Science* 282: 935–938.

# INDEX